Architectural Drafting and Design

Fourth Edition

Workbook

Alan Jefferis

David A. Madsen

Africa • Australia • Canada • Denmark • Japan • Mexico • New Zealand • Philippines
Puerto Rico • Singapore • Spain • United Kingdom • United States

Delmar Staff

Business Unit Director: Alar Elken
Executive Editor: Sandy Clark
Acquisitions Editor: Michael Kopf
Developmental Editor: John Fisher
Editorial Assistant:: Jasmine Hartman
Executive Marketing Manager: Maura Theriault
Channel Manager: Mona Caron

Marketing Coordinator: Paula Collins
Executive Production Manager: Mary Ellen Black
Production Manager: Larry Main
Art Director: Mary Beth Vought
Technology Project Manager: Tom Smith

COPYRIGHT © 2001, 2004
By Delmar Publishers
a division of International Thomson Learning Publishing Inc.

The ITP logo is a trademark under license.

Printed in the United States of America
4 5 6 7 8 9 10 XXX 05 04 03

For more information, contact Delmar, 5 Maxwell Dr., P.O. Box 8007, Clifton Park, NY 12065-8007; or find us on the World Wide Web at http://www.delmar.com

Asia:
Thomson Learning
60 Albert Street, #15-01
Albert Complex
Singapore 189969
Tel: 65 336 6411
Fax: 65 336 7411

Japan:
Thomson Learning
Palaceside Building 5F
I - I - I Hitotsubashi, Chiyoda-ku
Tokyo 100 0003 Japan
Tel: 813 5218 6544
Fax: 813 5218 6551

Australia/New Zealand:
Nelson/Thomson Learning
102 Dodds Street
South Melbourne, Victoria 3205
Australia
Tel: 61 39685 4111
Fax: 61 39 685 4199

UK/Europe/Middle East
Thomson Learning
Berkshire House
168-173 High Holborn
London
WC IV 7AA United Kingdom
Tel: 44 171497 1422
Fax: 44 171497 1426

Thomas Nelson & Sons LTD
Nelson House
Mayfield Road
Walton-on-Thames
KT 12 5PL United Kingdom
Tel: 44 1932 2522111
Fax: 44 1932 246574

Latin America:
Thomson Learning
Seneca, 53
Colonia Polanco
11560 Mexico D.F. Mexico
Tel: 525-281-2906
Fax: 525-281-2656

Canada:
Nelson/Thomson Learning
1120 Birchmount Road
Scarborough, Ontario
Canada MlK 5G4
Tel: 416-752-9100
Fax: 416-752-8102

Spain:
Thomson Learning
Calle Magallanes, 25
28015-Madrid
España
Tel: 34 91446 33 50
Fax: 34 91445 62 18

International Headquarters:
Thomson Learning
International Division
290 Harbor Drive, 2nd Floor
Stamford, CT 06902-7477
Tel: 203-969-8700
Fax: 203-969-8751

Library of Congress Cataloging-in-Publication Data
ISBN: 0-7668-1548–X

CONTENTS

SECTION 1

Basic Residential Projects

Note:
All directions for CAD projects are written assuming the use of AutoCAD 2000. Students using older versions of AutoCAD will also be able to complete these assignments. Although other CADD software can be used, some command names may vary. Electronic projects can be accessed through the following URL: www.cadd-drafting.com/olcs/jefferis/index.html

CHAPTER 1

Drawing Fundamentals

PROBLEM 1-1

Draw only the letters "PLAN" so that they will fit on 8 1/2 x 11"vellum. Use the architect's 16 scale (1"= 1") drafting machine or parallel bar and triangles, compass or circle template. If CAD is used, draw the letters at full scale. Make all lines of the letters dark and thick. Do not draw the dimensions.

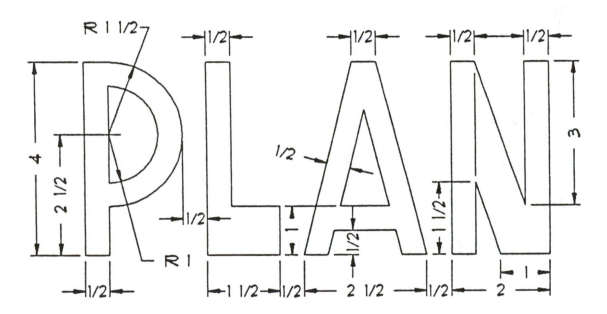

PROBLEM 1-2

Determine the length of the lines below using the architect's, engineers, and metric scales specified above each line. Place your answer in the blank following each scale identification.

Scale 1/4''=1'-0'' _____   1''=10'-0'' _____   1:1 _____

Scale 3/8''=1'-0'' _____   1''=20'-0'' _____   1:2 _____

Scale 1/2''=1'-0'' _____   1''=30'-0'' _____   1:5 _____

Scale 3/4''=1'-0'' _____   1''=40'-0'' _____   1:25 _____

Scale 1''=1'-0'' _____   1''=50'-0'' _____   1:50 _____

Scale 1/8''=1'-0'' _____   1''=100'-0'' _____   1:75 _____

Scale 1 1/2''=1'-0'' _____ 1''=60'-0'' _____   1:50 _____

Scale 3''=1'-0'' _____   1''=200'-0'' _____   1:5 _____

PROBLEM 1-3

Sketch straight lines between the numbered points below.

4
·

3 · · 2   6 · · 5

10 ·   · 9

1 ·   ·   ·   · 7
12   11   8

PROBLEM 1-4

Sketch the circles, given the center points and the radius of each circle below. For example, "C1" represents the center, and "R1" is a point establishing the radius for circle 1, "C2" and "R2" for circle 2.

R1   C1

R2   C2   R3   C3

C4   R4

PROBLEM 1-5

Given the sketch below of a swimming pool, patio, and spa, resketch larger in the rectangular grid provided. Use the grid as a guide as discussed in this chapter.

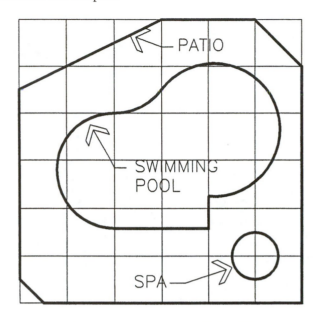

PROBLEM 1-6

Do the geometric construction exercises specified below.

A ──────────────────────── B

CONSTRUCT A PERPENDICULAR BISECTOR OF LINE AB ABOVE.

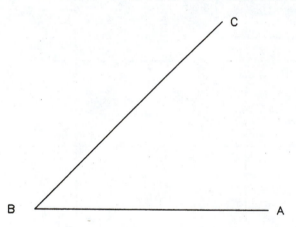

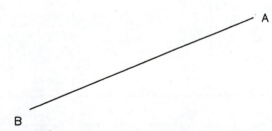

TRANSFER ANGLE ABC TO THE NEW LOCATION WITH LEG AB PLACED
ON THE GIVEN LINE AB.   BISECT ANGLE ABC AT THE NEW LOCATION.

PROBLEM 1-7

Do the geometric construction exercises specified below.

A ———————————————————————————————— B

DIVIDE LINE AB ABOVE INTO 12 EQUAL PARTS.

A ———————————————————————————————— B

C ———————————————————————————————— D

DIVIDE THE SPACE BETWEEN THE TWO LINES AB AND CD ABOVE INTO 14 EQUAL PARTS.

PROBLEM 1-8

Do the geometric construction exercises specified below.

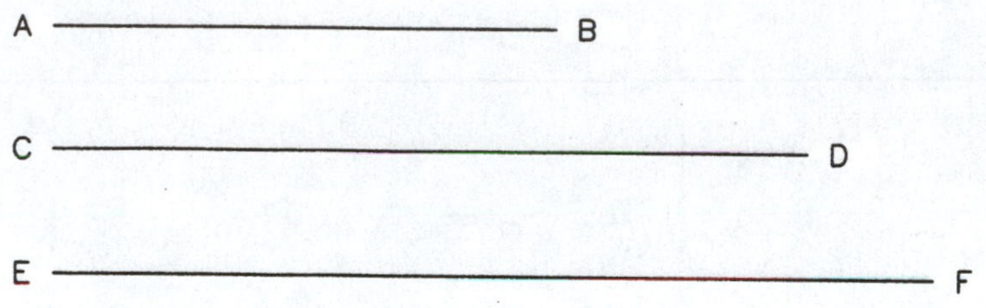

A ——————————————— B

C ——————————————— D

E ——————————————— F

CONSTRUCT A TRIANGLE IN THE SPACE PROVIDED ABOVE USING THE GIVEN
LINES AB, CD, AND EF AS THE SIDES OF THE TRIANGLE.

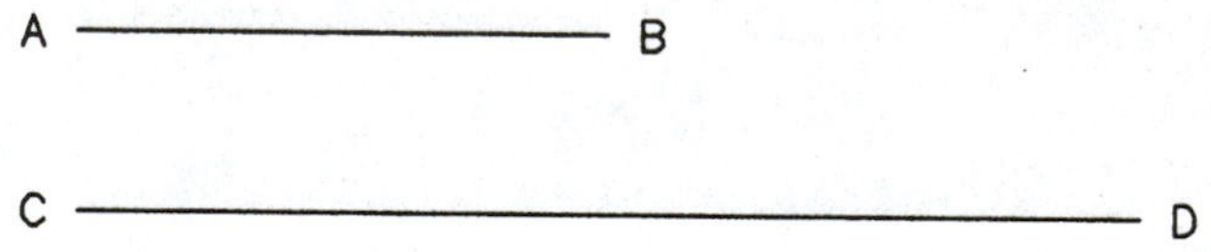

A ————————— B

C ——————————————— D

CONSTRUCT A RIGHT TRIANGLE IN THE SPACE PROVIDED ABOVE USING THE
GIVEN LINES AB AND CD AS THE SIDES OF THE RIGHT TRIANGLE.

PROBLEM 1-9

Given the top and side views below, complete by sketching the front views in the spaces provided.

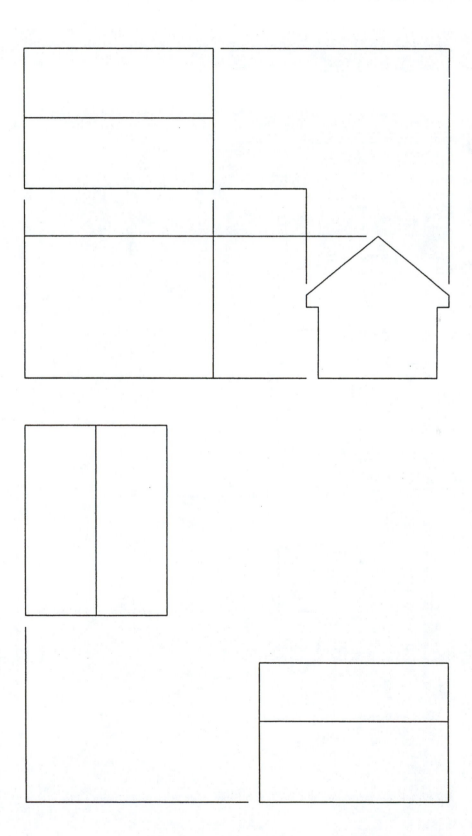

PROBLEM 1-10

Use proper sketching techniques to draw the objects below in the space provided. Use thin, dark lines for all objects.

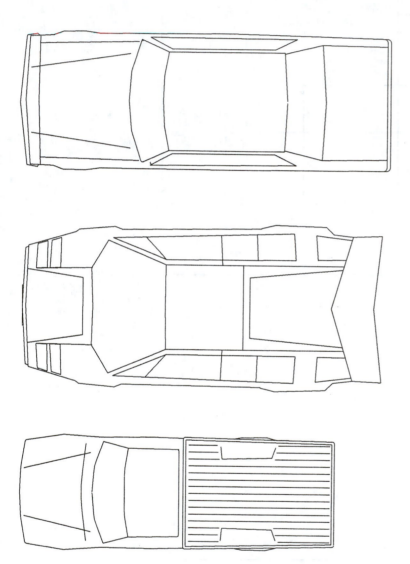

PROBLEM 1-11

Given the top and side views below, sketch the required front view in the space provided.

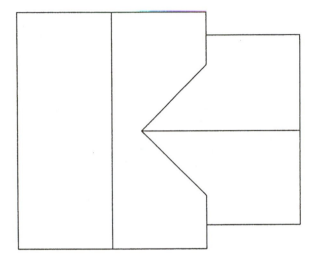

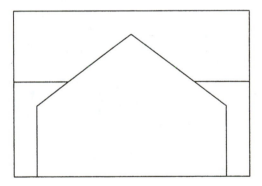

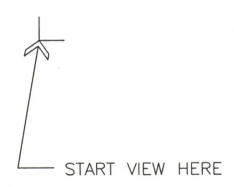

START VIEW HERE

PROBLEM 1-12

Approximate the dimensions from the given isometric drawing below to sketch the front, top, right-side, and left-side views in the space provided. The suggested front-view relationship is also provided.

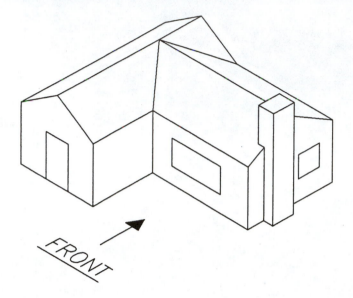

FRONT

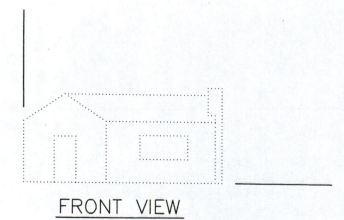

FRONT VIEW

PROBLEM 1-13

Approximate the dimensions from the given isometric drawing below to sketch the front, top, right-side, and left-side views in the space provided.

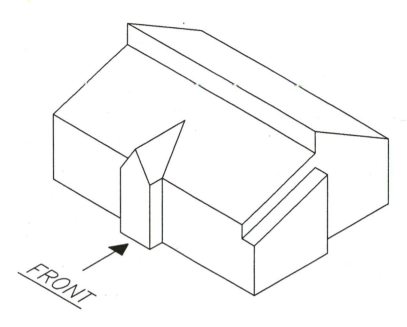

FRONT

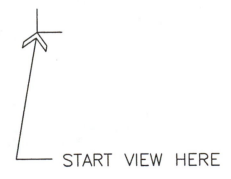

START VIEW HERE

PROBLEM 1-14

Given the multiviews of the objects below, prepare an isometric sketch in the location provided for each.

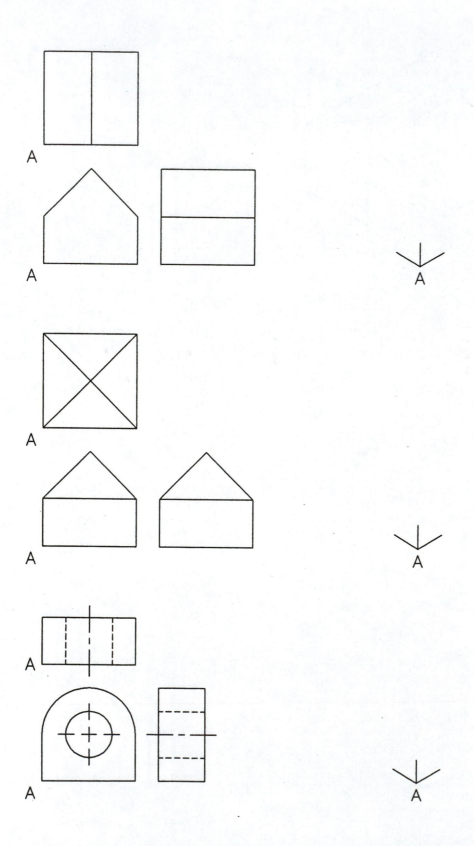

PROBLEM 1-15

Given the multiviews of the house below, prepare an isometric sketch in the space provided.

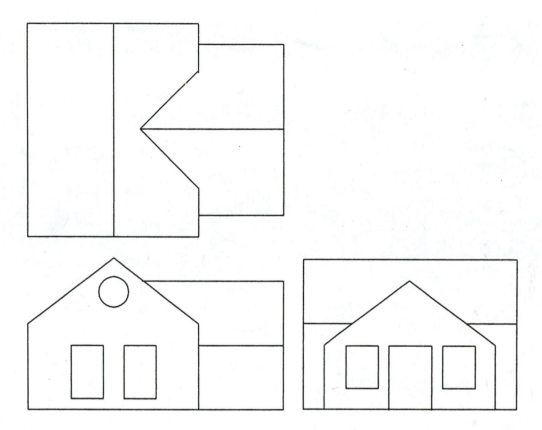

PROBLEM 1-16

This lettering shown below represents an architectural style. Duplicate each line of lettering on the guide lines provided. Practice making the letter shapes the same as the samples. The vertical guide lines help you keep your letters vertical.

PROBLEM 1-17

Duplicate the given lines of lettering on the guide lines provided below.

ARCHITECTURAL LETTERING IS
MORE INDIVIDUALIZED AND
ARTISTIC THAN LETTERING FOR
MECHANICAL DRAFTING. AS A
BEGINNING DRAFTER IT IS BEST
TO KEEP YOUR LETTERING STYLE
CONSERVATIVE. TOO MUCH FLAIR
MAY CAUSE YOUR LETTERING TO
LOOK UNNATURAL.

ALWAYS USE GUIDE LINES WHEN
DOING LETTERING. GUIDELINES
ARE SPACED APART EQUAL TO
THE HEIGHT OF THE LETTERS AND
ARE VERY LIGHTLY DRAWN.

PROBLEM 1-18

Duplicate the given lines of lettering on the guide lines provided below.

MINIMUM WINDOW AREA IS TO BE 1/10 FLOOR
AREA WITH NOT LESS THAN 10 SQ FT FOR
HABITABLE ROOMS AND 3 SQ FT FOR BATH-
ROOMS AND LAUNDRY ROOMS. NOT LESS
THAN ONE-HALF OF THIS REQUIRED WINDOW
AREA IS TO BE OPENABLE. EVERY SLEEP-
ING ROOM IS REQUIRED TO HAVE A WINDOW
OR DOOR FOR EMERGENCY EXIT. WINDOWS
WITH AN OPENABLE AREA OF NOT LESS THAN
5 SQ FT WITH NO DIMENSION LESS THAN 22 IN.
MEET THIS REQUIREMENT, AND THE SILL HEIGHT
IS TO BE NOT MORE THAN 44 IN. ABOVE THE
FLOOR.

CONCRETE MIX IS TO HAVE A MINIMUM ULTIMATE
COMPRESSIVE STRENGTH OF 2000 PSI AT 28
DAYS AND SHALL BE COMPOSED OF 1 PART
CEMENT, 3 PARTS SAND, 4 PARTS OF 1 IN. MAX.
SIZE ROCK, AND NOT MORE THAN 7 1/2 GALLONS
OF WATER PER SACK OF CEMENT.

PROBLEM 1-19

Duplicate each line of lettering on the guide lines provided. Practice making the letter shapes the same as the samples. The vertical guide lines help you keep your letters vertical.

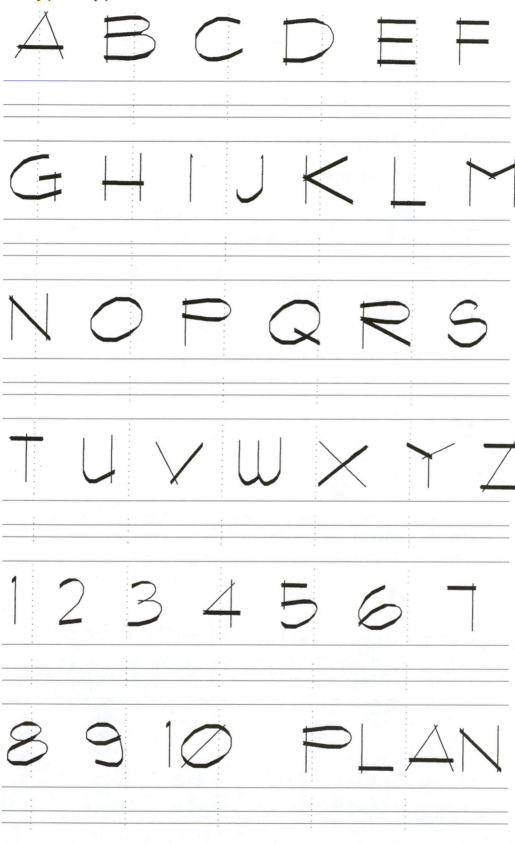

PROBLEM 1-20

Duplicate the given lines of lettering on the guide lines provided below.

FLOOR   PLAN   SITE

FLOOR PLAN SITE

FOUNDATION PLAN

FOUNDATION PLAN

ROOF   PLAN   CLOSET

ROOF PLAN CLOSET

LIVING ROOM   BATH

KITCHEN   BEDROOM

FAMILY ROOM UTILITY

SCALE: 1"=1'-Ø"   4X12 BEAM

WARDROBE   FOYER ENTRY

SCALE: 1/4"=1'-Ø"   6X14 BEAM

CHAPTER 2

Site Plan Projects

DIRECTIONS:

Draw one or more of the following site plans showing all property lines, setback lines and access roads. Use a scale suitable for clearly showing the site plan and all required material. Each site plan should include:

- Property and setback lines.
- Building footprint of one of the floor plans from Chapter 4.
- Legal description and summary of the project including the lot size and the structure size.
- North arrow.
- Access roads and easements.
- Driveways, parking areas, walkways, patios, and decks.
- Elevations of property corners and residence finish floor.
- Water, sewer, and power lines.
- Dimension to locate all setbacks, structures and utilities.
- Title and scale.

Unless specified, assume a front setback of 20'–0", a rear setback of 30'–0", one side yard of 5'–0" and the other side yard of 7'–0". Contact your local zoning department for sites that are not part of the Barrington Heights subdivision to obtain the setbacks for sites of similar size in your area.

Assume the sites in Problems 2–8 through 2–13 will have septic systems and private water supply. Provide a 1200-gallon septic tank, 600' of distribution lines, space for 300' of distribution lines for future expansion, and a well. Maintain the following dimensions:

well to house —10' min.	septic tank to house —10' min.
well to drain field — 100'	drain lines to property line —10' min.
between drain lines — 10' min.	maximum drain line length —125'

PROBLEM 2-1

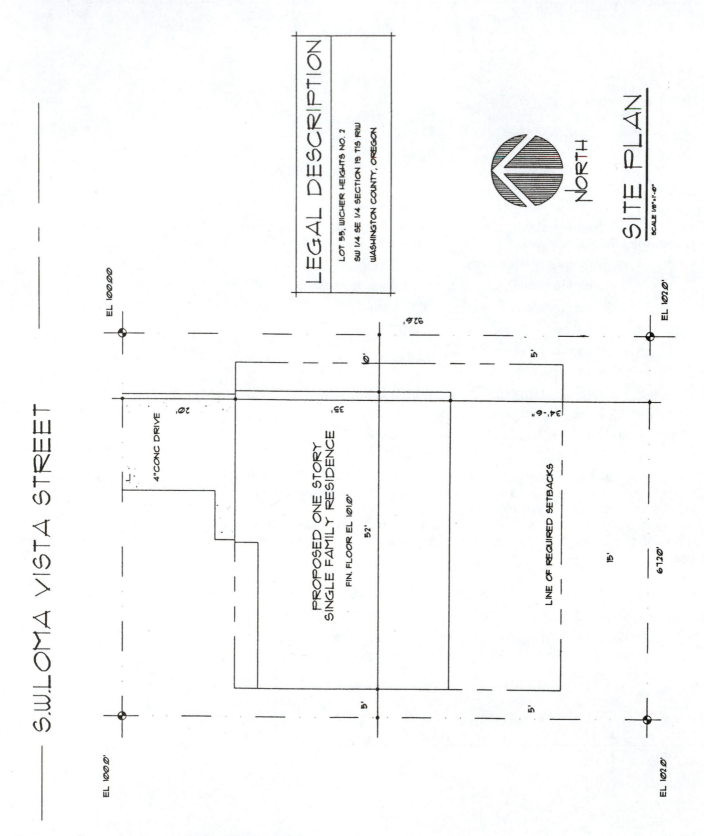

LEGAL DESCRIPTION

LOT 55, WICHER HEIGHTS NO. 2
SW 1/4 SE 1/4 SECTION 19 T15 R1W
WASHINGTON COUNTY, OREGON

SITE PLAN

SCALE 1/8"=1'-0"

NORTH

S.W. LOMA VISTA STREET

4" CONC DRIVE

PROPOSED ONE STORY
SINGLE FAMILY RESIDENCE

FIN. FLOOR EL 101.0'

LINE OF REQUIRED SETBACKS

EL 100.0'

EL 102.0'

Courtesy of Sunridge Design, Wally Greiner AIBD.

PROBLEM 2-2

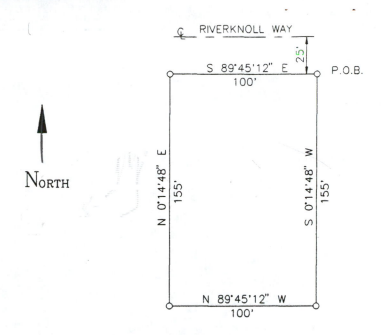

CL RIVERKNOLL WAY

S 89°45'12" E
100'

25'

P.O.B.

N 0°14'48" E
155'

S 0°14'48" W
155'

N 89°45'12" W
100'

LOT 12
BLOCK 1
BARRINGTON HEIGHTS
YOUR CITY, COUNTY
STATE

PROBLEM 2-3

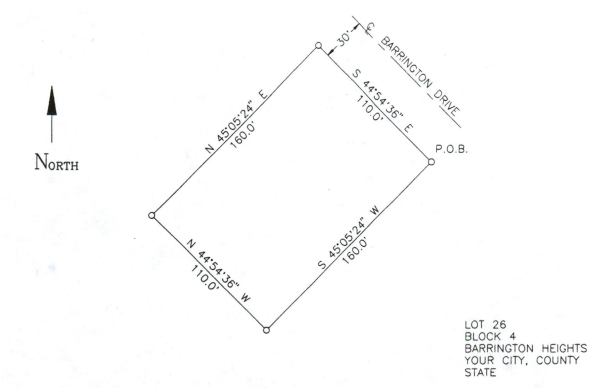

30'

CL BARRINGTON DRIVE

N 45°05'24" E
160.0'

S 44°54'36" E
110.0'

P.O.B.

N 44°54'36" W
110.0'

S 45°05'24" W
160.0'

LOT 26
BLOCK 4
BARRINGTON HEIGHTS
YOUR CITY, COUNTY
STATE

PROBLEM 2-4

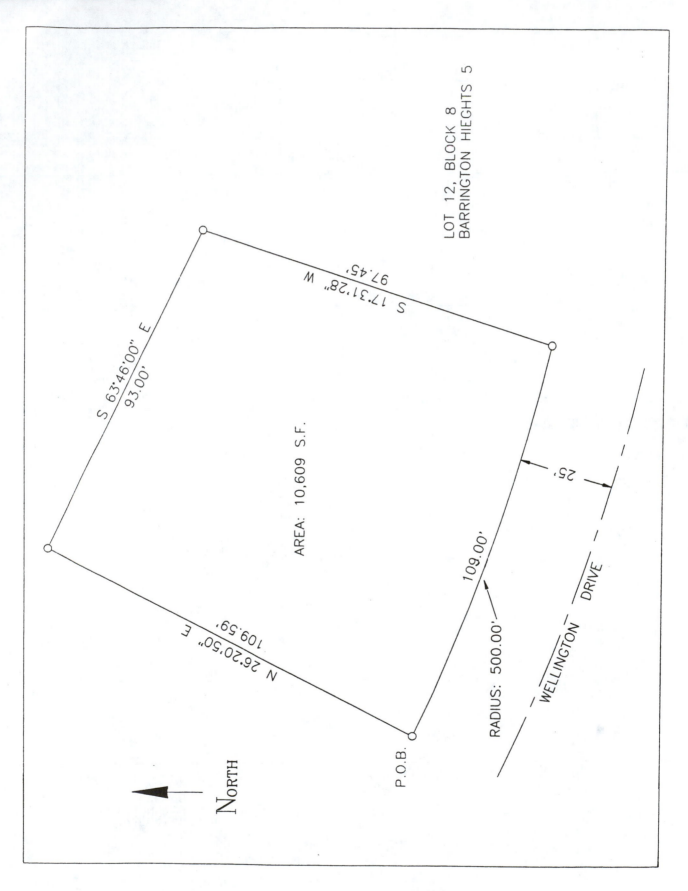

LOT 12, BLOCK 8
BARRINGTON HIEGHTS 5

S 63°46'00" E
93.00'

S 17°31'28" W
97.45'

AREA: 10,609 S.F.

25'

109.00'

N 26°20'50" E
109.59'

RADIUS: 500.00'

WELLINGTON DRIVE

P.O.B.

NORTH

PROBLEM 2-5

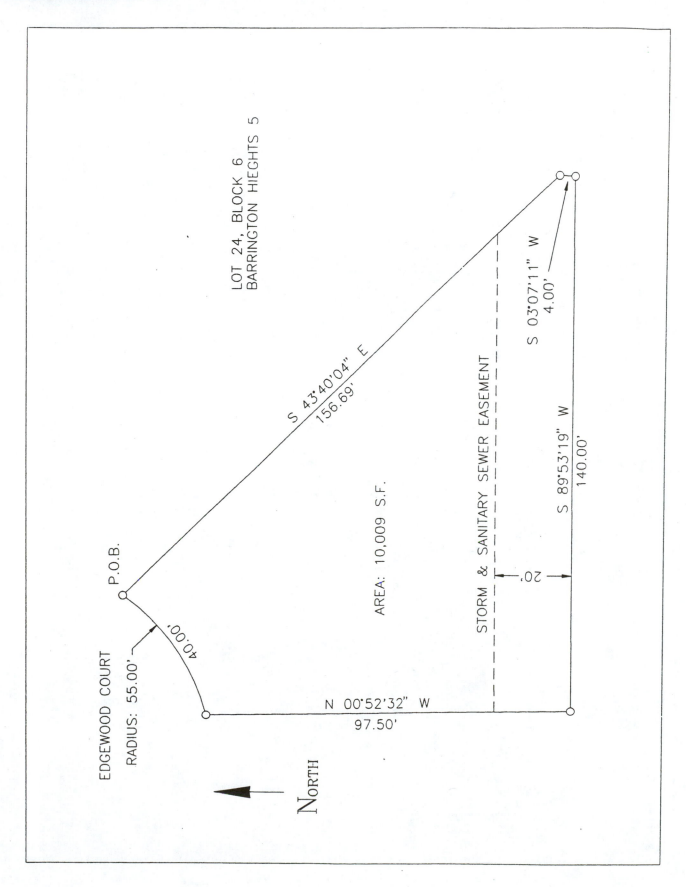

PROBLEM 2-6

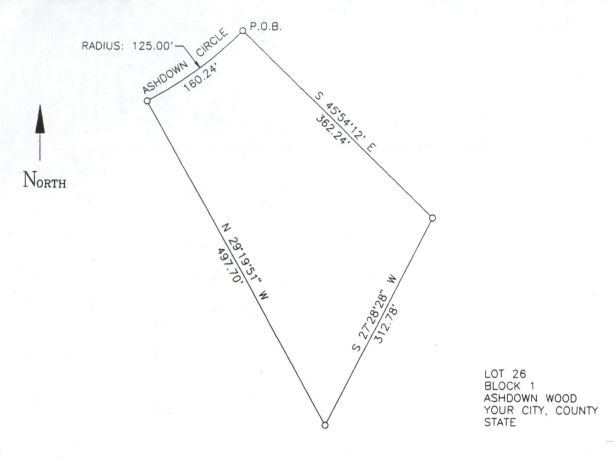

RADIUS: 125.00'

ASHDOWN CIRCLE
160.24'

P.O.B.

S 45°54'12' E
362.24'

North

N 29°19'51" W
497.70'

S 27°28'28" W
312.78'

LOT 26
BLOCK 1
ASHDOWN WOOD
YOUR CITY, COUNTY
STATE

PROBLEM 2-7

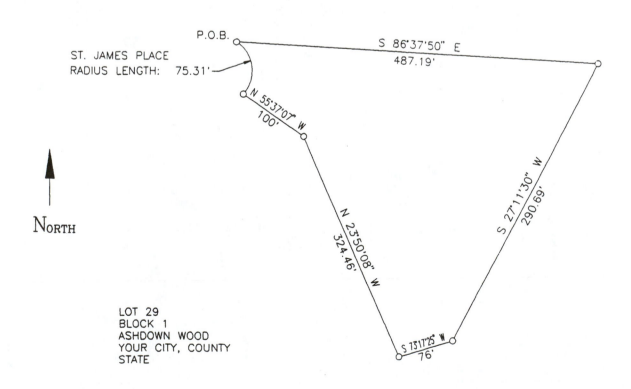

P.O.B.

ST. JAMES PLACE
RADIUS LENGTH: 75.31'

S 86°37'50" E
487.19'

N 55°37'07" W
100'

North

N 23°50'08" W
324.46'

S 27°11'30" W
290.69'

S 73°17'25" W
76'

LOT 29
BLOCK 1
ASHDOWN WOOD
YOUR CITY, COUNTY
STATE

PROBLEM 2-8

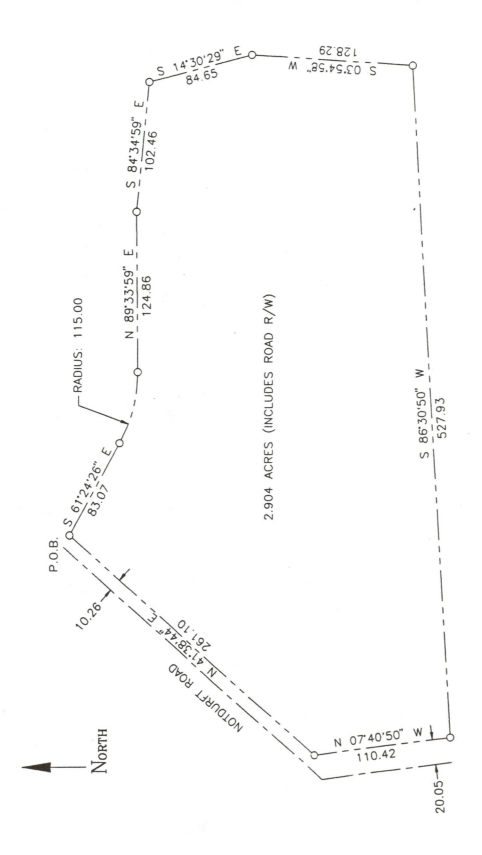

PROBLEM 2-9

Given the plot plan boundaries below, place the house using the minimum property line setbacks of 25' front, 25' back, 7' sides. Draw the driveway, patios, walks, utilities, and other items to your own specifications.

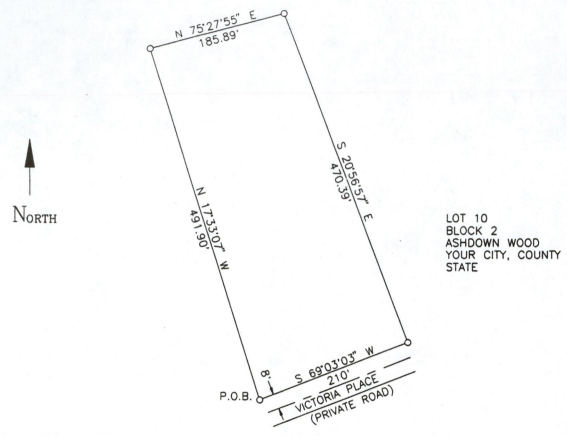

LOT 10
BLOCK 2
ASHDOWN WOOD
YOUR CITY, COUNTY
STATE

PROBLEM 2-10

Given the plot plan boundaries below, place the house using the minimum property line setbacks of 30' front, 25' back, 7' on one side, aid 5' on the other side. Draw the driveway, patios, walks, utilities, and other items to your own specifications.

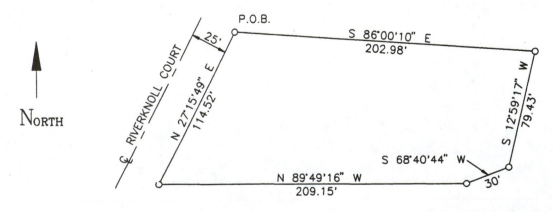

LOT 15
BLOCK 3
BARRINGTON HEIGHTS
YOUR CITY, COUNTY
STATE

Directions:
Use the following legal descriptions to layout the construction sites for Problems 2–11 through 2–13. Once the lot shape has been determined, complete a site plan by providing the required information to describe the site and the structure to be built.

PROBLEM 2-11

Victoria Place front street. Beginning at the southeast property corner thence along Victoria Place 210' S69°03'03"W thence 491.90' N17°33'07"W thence 185.89' N75°27'55"E thence 470.39' N20°56'57"W to the point of beginning.

PROBLEM 2-12

Ashdown Circle front street. Beginning at the northeast corner thence 497.70' N29°19'51" thence 297.13' S86°37'50"E thence 369.13' N14°45'49"W thence along Ashdown Circle on an arc 158.90' long back to the point of beginning.

PROBLEM 2-13

Waterford Place front street. Beginning at the northwest corner thence 110.78' N63°06'53"E thence 211.61' S87°28'51"E thence 295.61' S18°31'14"E thence 255' S62°47'35"W thence 234.88' N18°31'14"W thence 111.15' N65°19'W thence 90' on an arc back to the point of beginning.

Directions:
Use the site plan created in Problems 2–2 and 2–3 to complete
Problem 2–14 and 2–15 respectively. Use the attached engineer's field
notes to layout the ground elevations for each site. Once the elevation
locations have been determined, layout contour lines to represent 1'
intervals. Draw all indicated contour lines and highlight lines at 5'
intervals.

PROBLEM 2-14

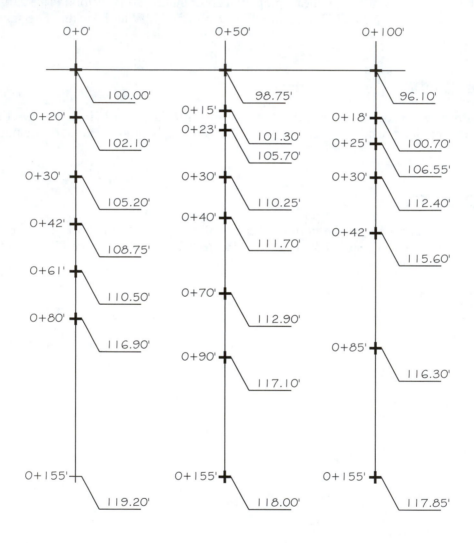

PROBLEM 2-15

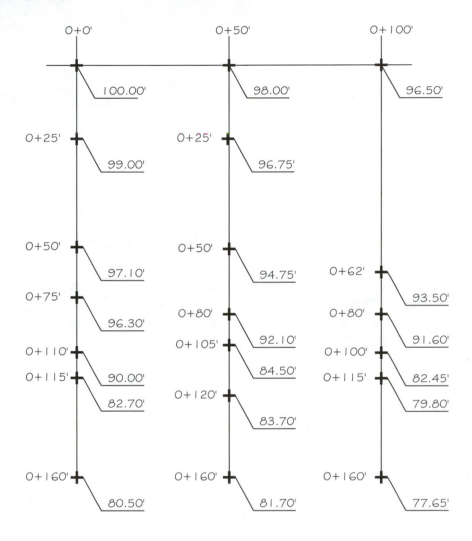

CHAPTER

Floor Plan Fundamentals

PROBLEM 3-1

Given the wall layout for a kitchen below, draw the base and upper cabinets and the appliances specified at the locations shown.

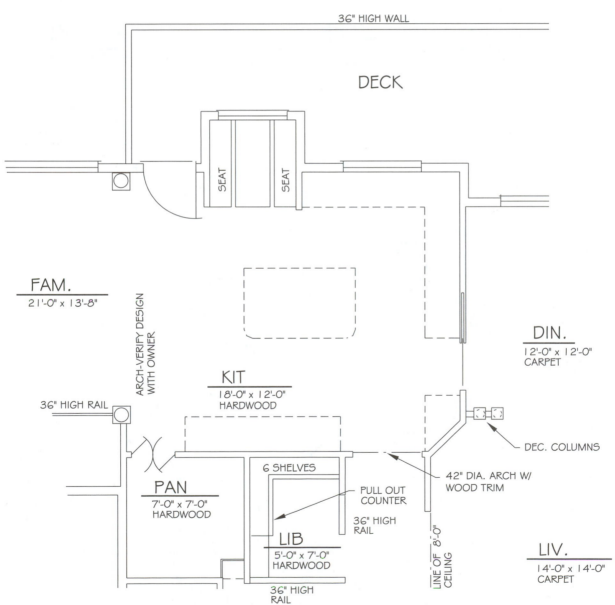

Courtesy of Alan Mascord Design Associates.

PROBLEM 3-2

Given the wall layouts for the bathrooms below, draw the plumbing fixtures specified at the locations shown.

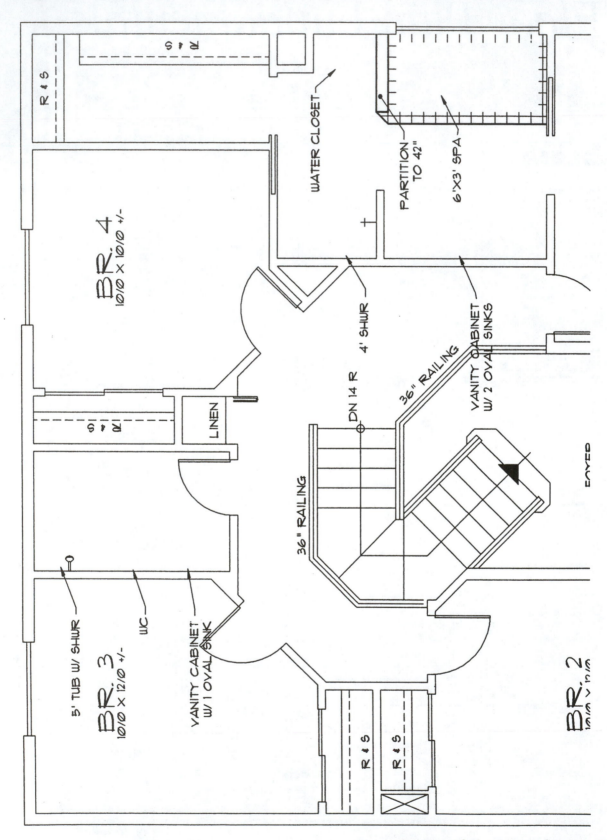

Courtesy of Alan Mascord Design Associates.

PROBLEM 3-3

Given the wall layout for the utility and bathroom below, draw the items specified at the specified location.

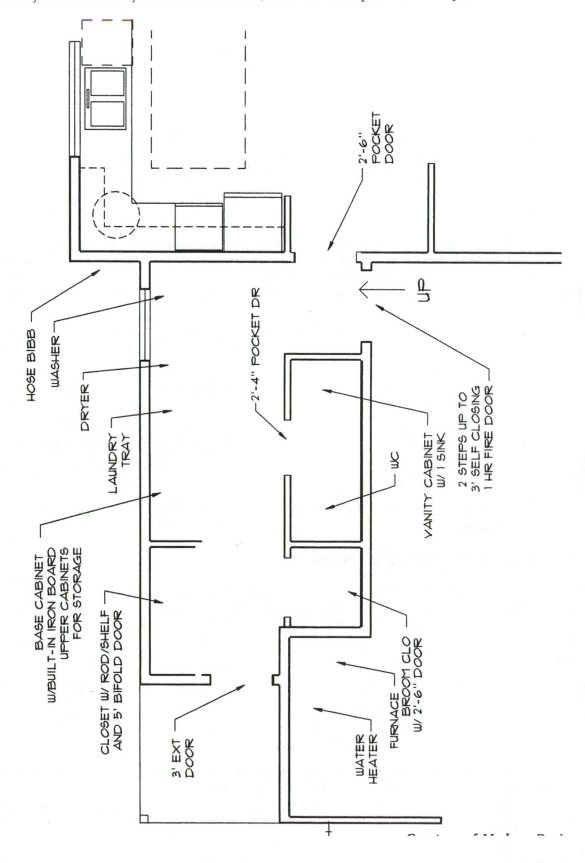

HOSE BIBB

WASHER

DRYER

LAUNDRY TRAY

BASE CABINET W/BUILT-IN IRON BOARD UPPER CABINETS FOR STORAGE

CLOSET W/ ROD/SHELF AND 5' BIFOLD DOOR

3' EXT DOOR

2'-4" POCKET DR

2'-6" POCKET DOOR

UP

2 STEPS UP TO 3' SELF CLOSING 1 HR FIRE DOOR

VANITY CABINET W/ 1 SINK

WC

WATER HEATER

FURNACE

BROOM CLO W/ 2'-6" DOOR

PROBLEM 3-4

Draw a floor plan and show the following using 1/4"=1'-0" scale. Note: The washer and dryer are in the construction contract unless otherwise specified.

- Clothes washer, dryer, and laundry tray along a wall segment with an upper cabinet for storage.
- Clothes washer and dryer (not in construction contract) in a closet with shelf above and bi-fold doors.
- Built-in and surface-mount ironing board.
- Furnace and water heater in closet with bi-fold doors.
- Bathroom vanity with lavatory providing laundry chute to utility below and line with 26-ga. metal.
- Clothes washer and dryer with cabinet above and laundry chute from cabinet above.
- Wardrobe closet with shelf and pole, and 6'-0 wide bipass doors.

PROBLEM 3-5

Use CAD to create the following symbols for a floor plan. Draw each object at full size. Create each symbol using the sizes and linetypes used in Chapter 14 of the text. If text is required to identify the symbol, use a height suitable for display at 1/4" = 1'-0". Save each symbol as a separate WBLOCK using a name that will describe the object. Establish folders and sub folders to store these drawings. Use the title Floor Plan Prototypes for the folder name, the subheading listed below for subfolder names and the object name for the drawing name.

Doors: 36" exterior swinging door, 32" exterior swinging door, an exterior pair of 30" swinging doors, 6' sliding door, 32" interior swinging door, 30" interior swinging door, 28" interior swinging door.

Window: *Draw a 12" wide window suitable for insertion in a 6" wide wall. Draw a second window suitable for insertion in a 4" wide wall. Note: remember the X scale factor can be used to alter the width of the window block as it is inserted.*

Appliances: Draw the following appliances: Refrigerator, dishwasher, trash compactor, cooktop, free-standing stove, double oven, washer, dryer, 24"Ø water heater, 18" × 18" gas FAU, 30" Ø central vacuum.

Plumbing: 42" × 21" triple sink, 32" × 21" double sink, 16" vegetable sink, laundry tray, 19" × 16" oval lavatory, water closet, 60" tub, 48" × 36" shower, 42" × 42" shower, 36" × 36" shower.

Miscellaneous: 22" × 30" attic access, 22" × 48" attic access (with ladder), 24" × 24" skylight, 24" × 48" skylight, built–in ironing board.

PROBLEMS 3-6 THROUGH 3-14

Use the following guidelines to complete problems 3–6 through 3–14. If problems are completed manually use a separate sheet of vellum and a scale of 1/4" = 1'-0". If drawings are completed using CAD, use linetype and lettering scale factors suitable for plotting at a scale of 1/4" = 1'-0". Save each drawing as a separate file.

PROBLEM 3-6

Draw the following floor plan items and label the materials:

- Entry area similar to that shown in Chapter 14 with stone flooring.
- Entry area similar to that shown in Chapter 14 with brick flooring.
- Shower with tile.
- Letter the following room names and floor materials: LIVING (carpet), DINING (vinyl), BATH (tile), FOYER (marble), KITCHEN (hardwood).

PROBLEM 3-7

Draw a 14' x 14' sunken (one step) living room and a 12' x 12' raised (two steps) dining room.

PROBLEM 3-8

Draw the following masonry fireplace representations.

- Single
- Back-to-back double
- Corner
- Double-faced see-through
- Three-faced
- Swedish

PROBLEM 3-9

Draw the following manufactured fireplace or solid-fuel-burning appliance representations.

- Manufactured firebox framed in masonry.
- Manufactured firebox framed in wood.
- Wood stove on tile hearth and 5' high brick veneer wall protection.
- Corner wood burning stove with 12" raised stone hearth and stone wall protection.
- Masonry alcove for solid-fuel-burning stove.

PROBLEM 3-10

Draw the following floor plan masonry fireplace combinations.

- Fireplace with wood storage.
- Fireplace with clean out.
- Fireplace with gas outlet.
- Fireplace with adjacent barbecue unit.

PROBLEM 3-11

Use a scale of 1/4" = 1'-0" to design a utility room that includes a washer, dryer, sewing center, pantry, laundry sink in a base cabinet, and a floor drain.

PROBLEM 3-12

Draw the floor plan for each of the following kitchen types. Show a sink, range/oven, refrigerator, and dishwasher with each.

- U Shape
- Peninsula

PROBLEM 3-13

Draw the floor plan for each of the following kitchen types. Show a sink, range/oven, refrigerator, and dishwasher with each.

- One wall
- L shape

PROBLEM 3-14

Draw the floor plan for each of the following kitchen types. Show a sink, range/oven, refrigerator, and dishwasher with each.

- Island
- Corridor

PROBLEM 3-15A

This is a pencil problem. Given the screened image below of a partial floor plan, use the proper pencil techniques to darken all lines. Draw all wall outlines as thick lines and all other lines thin. Trace all lettering for practice in developing your architectural lettering style.

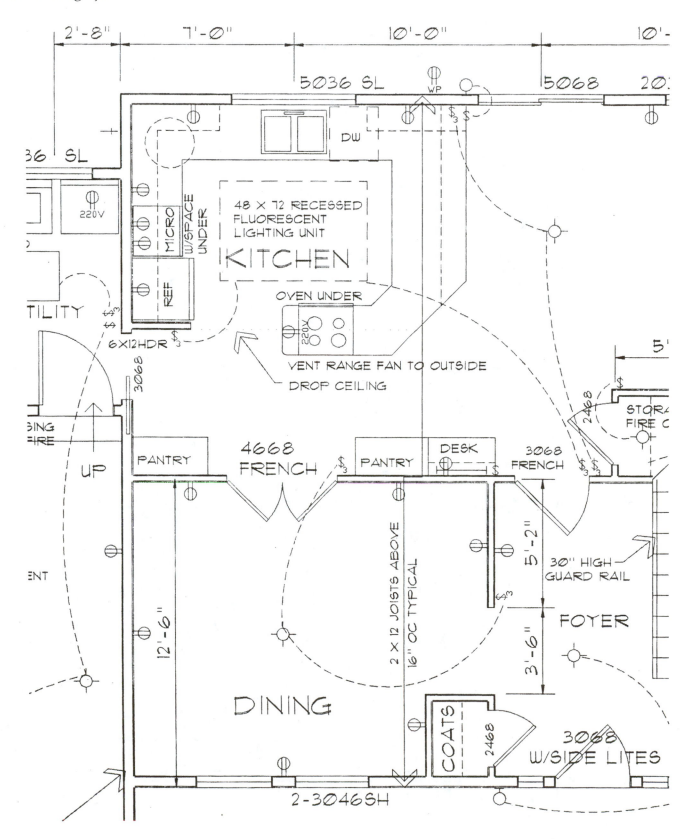

PROBLEM 3-15-B

Use the floor plan drawing from Problem 3-15A and create the drawing using CAD. Assign thick lines for all walls and thin lines for all other objects.

PROBLEM 3-16

This is an inking problem. Given the screened image below of a partial floor plan, use the proper technical pens and inking techniques to darken all lines. Draw all wall outlines as thick lines and all other lines thin. Omit all lettering and dimensions.

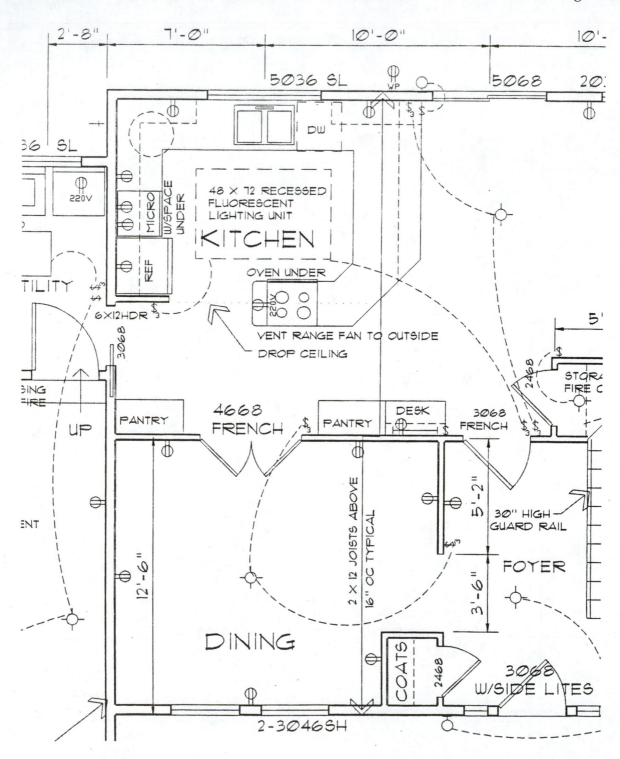

PROBLEM 3-17

Given the floor plan below, label the living room, family room, dining room, kitchen, nook, appliances, stairs, and equipment.

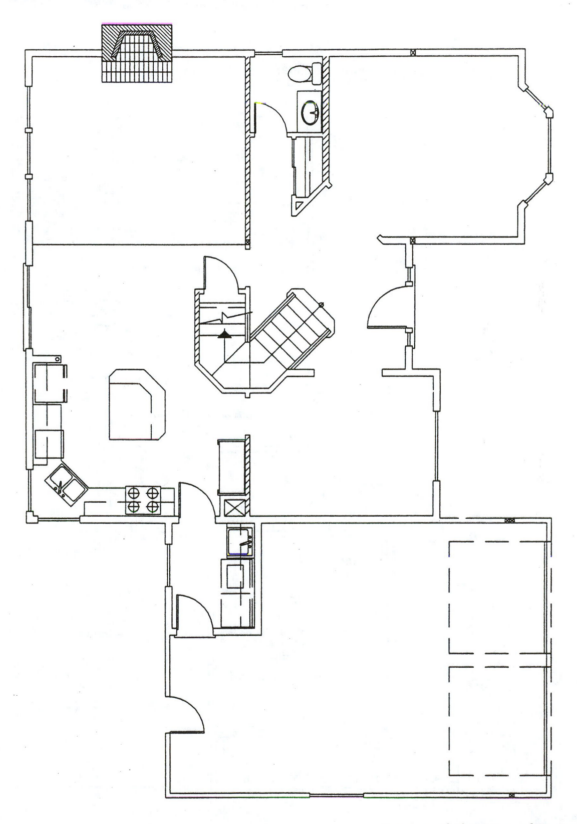

Courtesy of Alan Mascord Design Associates.

PROBLEM 3-18

Given the partial floor plan below, provide a number symbol for each door and a letter symbol for each window, and complete the door and window schedule on page 45 with the information specified for each.

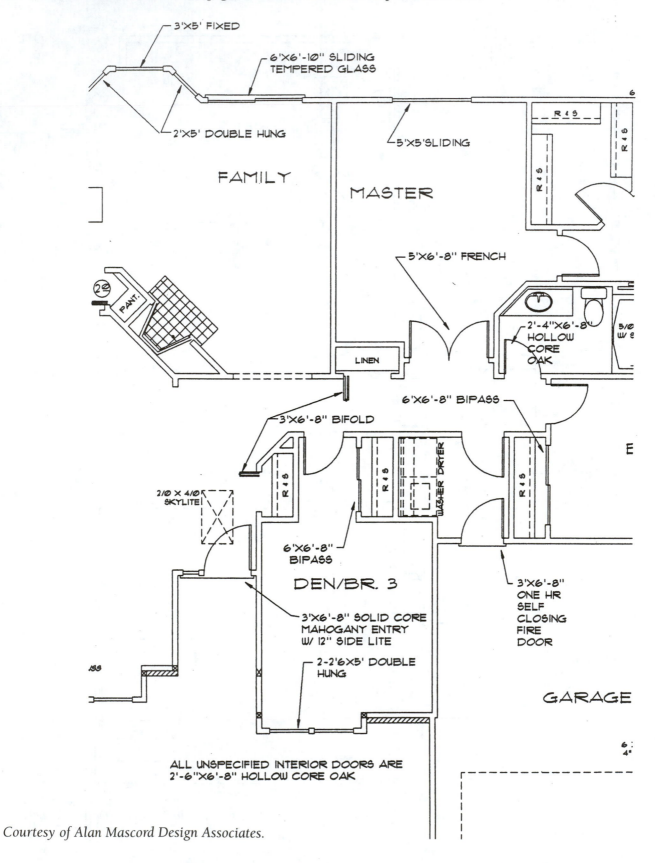

3'X5' FIXED

6'X6'-10" SLIDING TEMPERED GLASS

2'X5' DOUBLE HUNG

5'X5'SLIDING

FAMILY

MASTER

5'X6'-8" FRENCH

R 19

R 46

R 15

2'-4"X6'-8" HOLLOW CORE OAK

LINEN

6'X6'-8" BIPASS

3'X6'-8" BIFOLD

WASHER DRYER

R 46

R 15

R 46

2/0 X 4/0 SKYLITE

6'X6'-8" BIPASS

DEN/BR. 3

3'X6'-8" SOLID CORE MAHOGANY ENTRY W/ 12" SIDE LITE

3'X6'-8" ONE HR SELF CLOSING FIRE DOOR

2-2'6X5' DOUBLE HUNG

GARAGE

ALL UNSPECIFIED INTERIOR DOORS ARE 2'-6"X6'-8" HOLLOW CORE OAK

Courtesy of Alan Mascord Design Associates.

PROBLEM 3-18 (CONTINUED)

DOOR SCHEDULE			
	QTY	SIZE	TYPE

WINDOW SCHEDULE			
	QTY	SIZE	TYPE

PROBLEM 3-19

Label the components of each of the cross-section cutting plane symbols in the blank spaces provided.

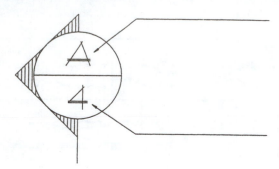

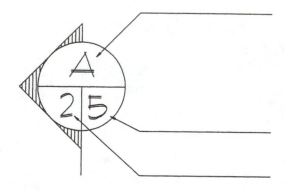

PROBLEM 3-20

Given the partial floor plan on page 47 place the following specific notes next to the corresponding numbers shown on the floor plan:

1. 2/8 x 6/8 s.c./s.c.
2. MODEL UGLC-OTEC-GS 80% EFFICIENT NAT. GAS FURN. W/WATER HEATER (PILOT ABOVE 18") W/3" 0 CONC. FILLED GUARD PIPE
3. GIRDER TRUSS
4. 2/8 x 6/8 s.c. 1-LITE (TEMP)
5. 1/2" GYP. BD. ON WALLS AND CLG.
6. 4-2 X 4
7. 4" CONC. SLAB W/6 x 6-10 WWM ON 4" MIN. GRANULAR FILL
8. 16/0 x 7/0 OVERHEAD GARAGE DR.
9. 4 x 14 HDR.
10. H.B.
11. 2-2 x 6
12. 2 x 6 CLG. JST. @ 24" OC
13. 2-2 x 4
14. 2-2 x 4
15. 4-2 x 4
16. 2-2/6 x 6/0 AL. S.H. (36" VENTS) (TEMP)
17. BRICK VENEER OVER 1" AIR SPACE W/26 GA. METAL TIES @ 24" OC @ EA. STUD OVER TYVEK.

PROBLEM 3-20 (CONTINUED)

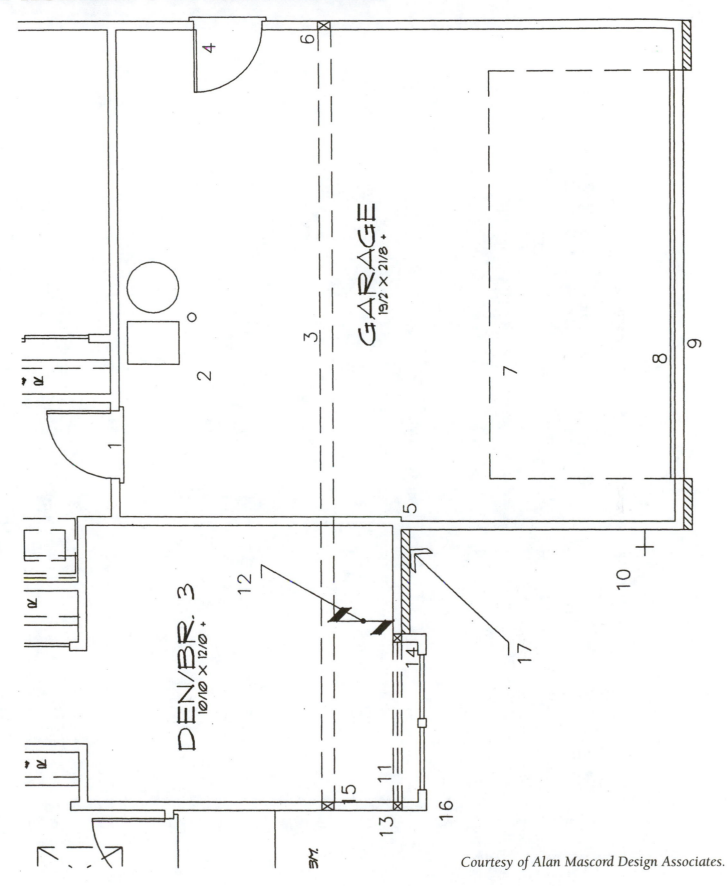

Courtesy of Alan Mascord Design Associates.

CHAPTER 4

Residential Floor Plan Projects

DIRECTIONS: The floor plan drawing problems are given in increasing difficulty. Use the following general information and instructions to complete the problems:

- Floor plan layouts with key dimensions are provided. Each layout can also be found at the Web site. Fewer and fewer dimensions are provided as problems increase in difficulty. When dimensions are omitted it is your responsibility to determine the appropriate measurements and provide the dimensions on your final drawings.

- Rooms are named where appropriate. If no name is provided, determine the best use for the room and provide a room name.

- A proposed electrical layout is given in some problems. Ignore this information until the electrical plan is completed. You may have to locate switches and provide other electrical installations as discussed in the main text. Advanced problems require you to design the elctrical layout.

- Use a 1/4" = 1'-0" scale.

- Use appropriately sized drawing sheets depending on the dimensions of the plan or your specific course objectives and instructions.

- Door and window sizes and types are given on some problems. Determine appropriate windows and doors when none are specified. Door and window sizes may be placed on the plan or placed in door and window schedules.

PROBLEM 4-1

One-story house.

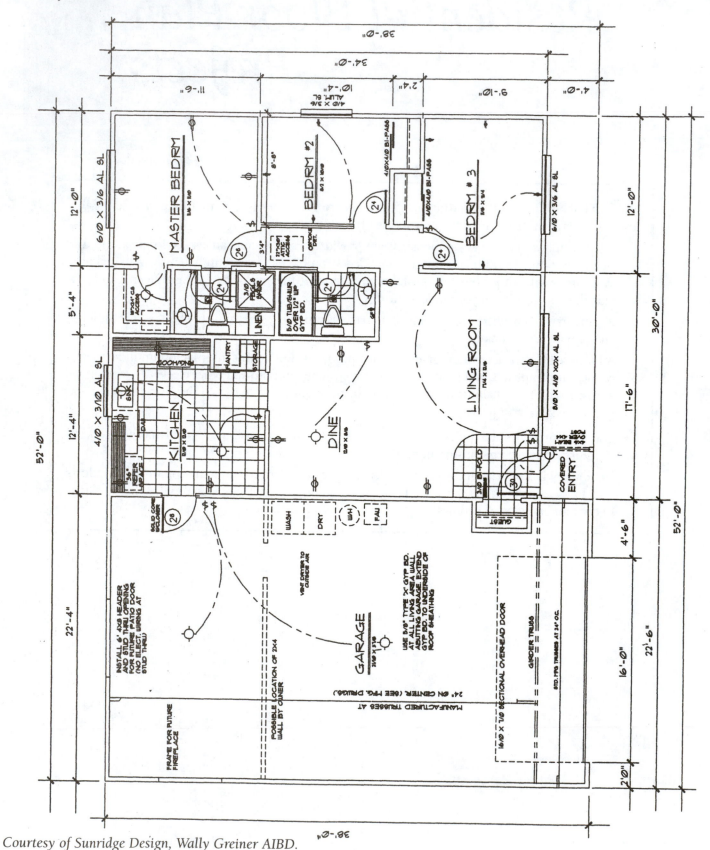

Courtesy of Sunridge Design, Wally Greiner AIBD.

PROBLEM 4-2

One-story house.

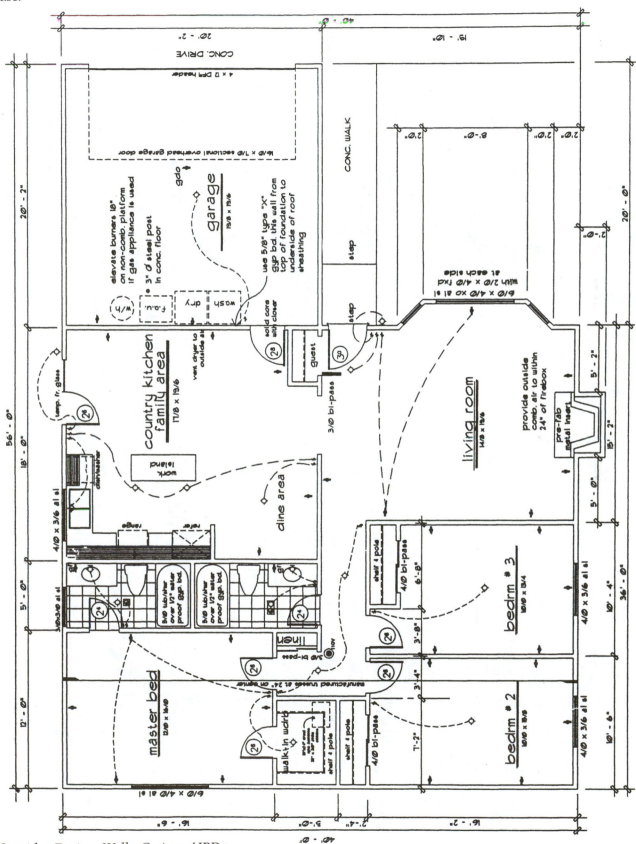

Courtesy of Sunridge Design, Wally Greiner AIBD.

PROBLEM 4-3

One-story house.

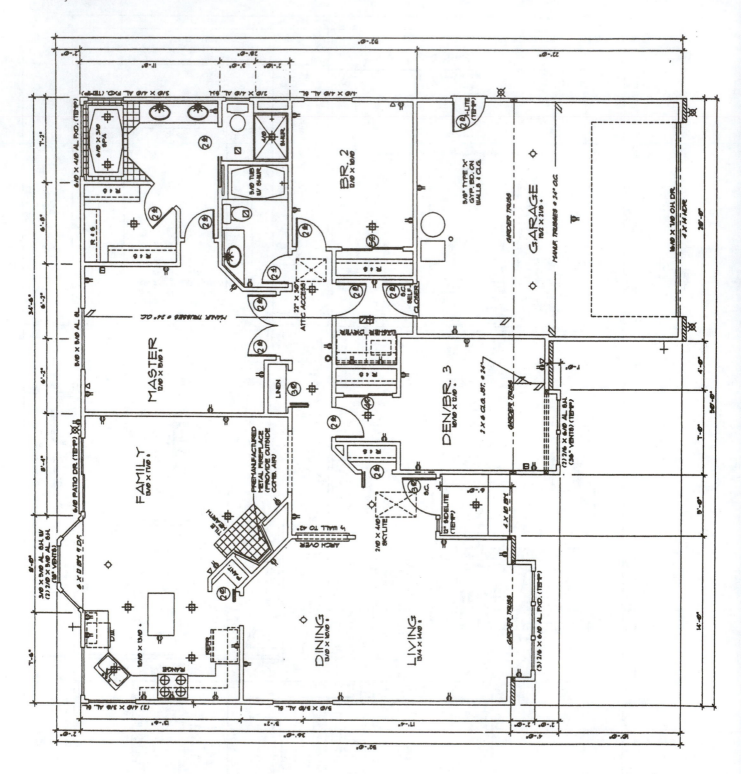

Courtesy of Alan Mascord Design Associates.

PROBLEM 4-4

One-story house.

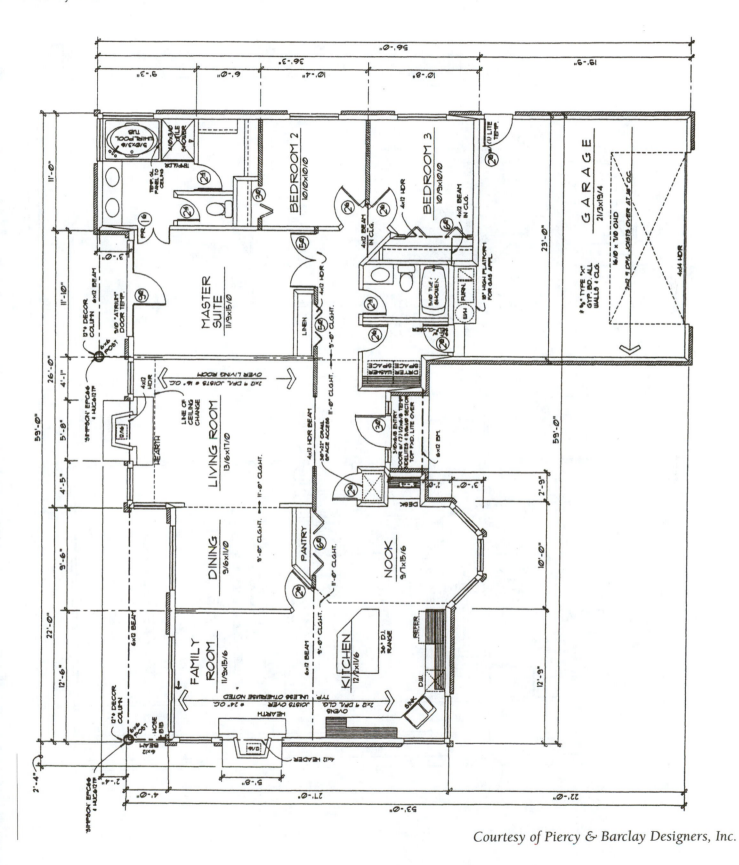

Courtesy of Piercy & Barclay Designers, Inc.

PROBLEM 4-5
One-story house

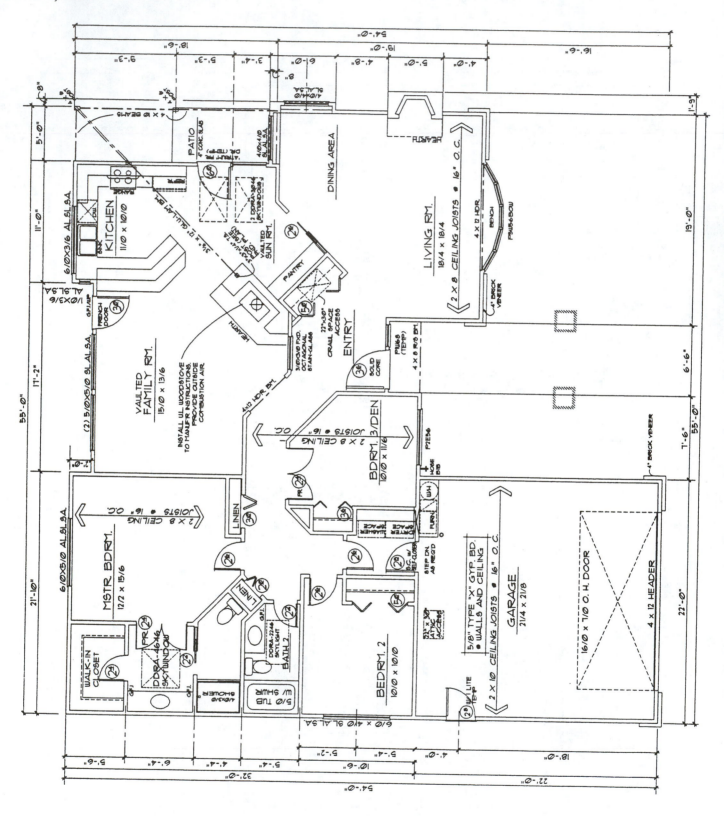

Courtesy of Piercy & Barclay Designers, Inc.

PROBLEM 4-6

One-story house

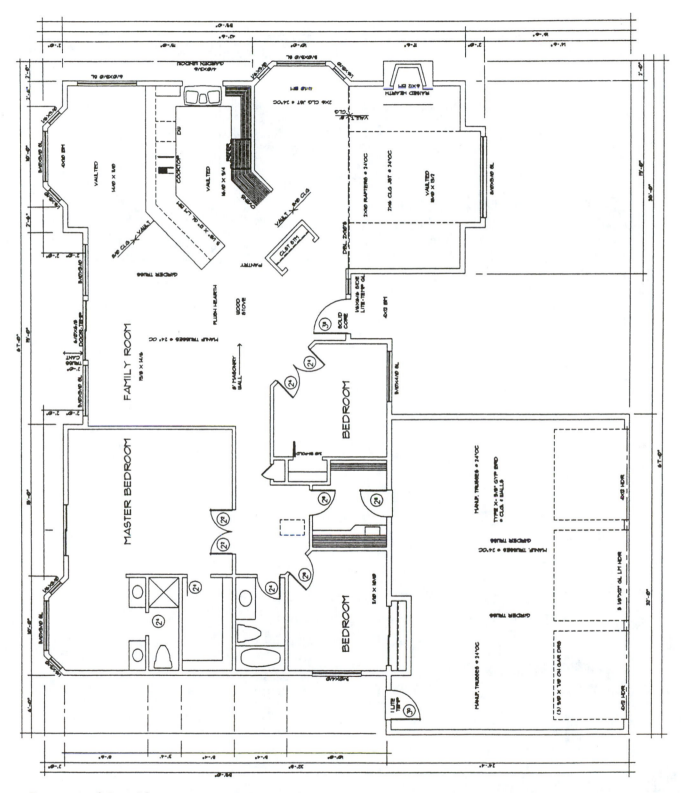

Courtesy of Sunridge Design, Wally Greiner AIBD.

PROBLEM 4-7

One-story house. Limited dimensions are provided and window sizes are omitted. Establish window sizes and types based on the code and window discussion in the main text.

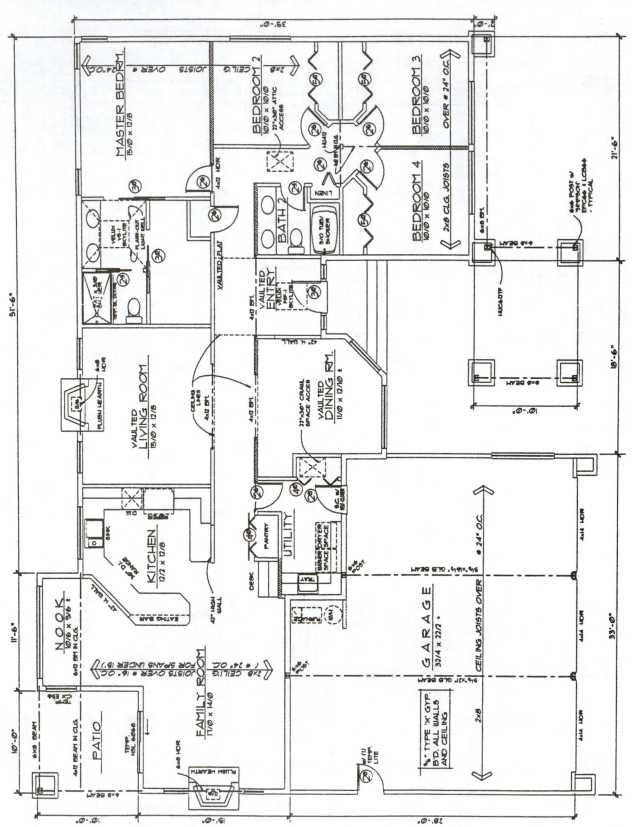

Courtesy of Piercy & Barclay Designers, Inc.

PROBLEM 4-8

This drawing problem provides you with the basic floor plan layout. Design and draw the floor plan, and design the door and window sizes to your own specifications, which should meet or exceed minimum code requirements.

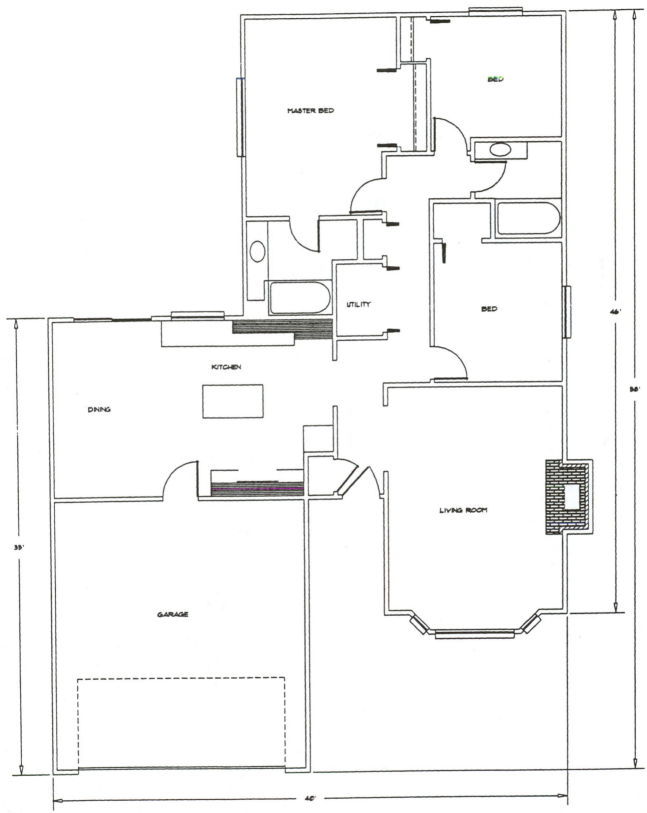

Courtesy of Sunridge Design, Wally Greiner AIBD.

PROBLEM 4-9

This drawing problem provides you with the basic floor plan layout. Design and draw the floor plan, and design the door and window sizes to your own specifications, which should meet or exceed minimum code requirements.

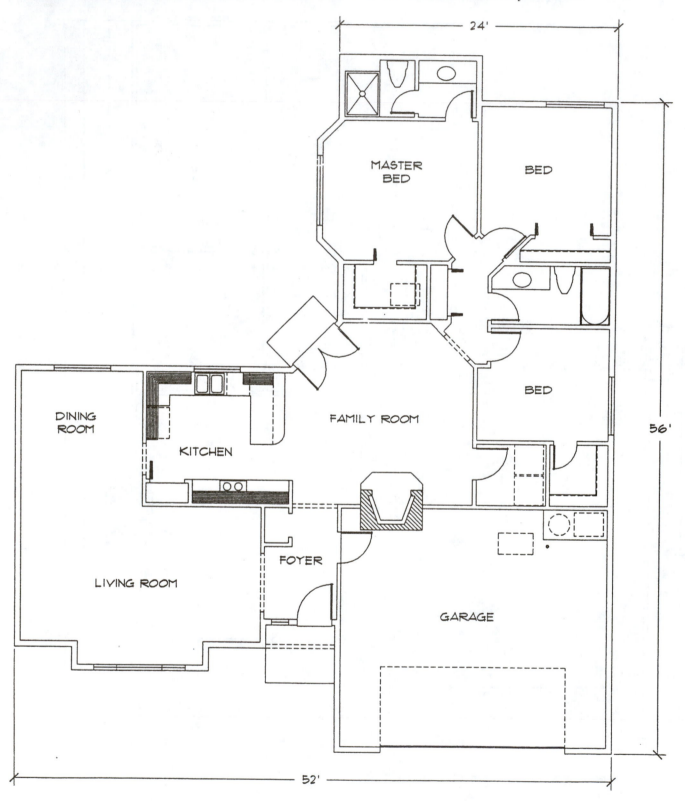

Courtesy of Sunridge Design, Wally Greiner AIBD.

PROBLEM 4-12

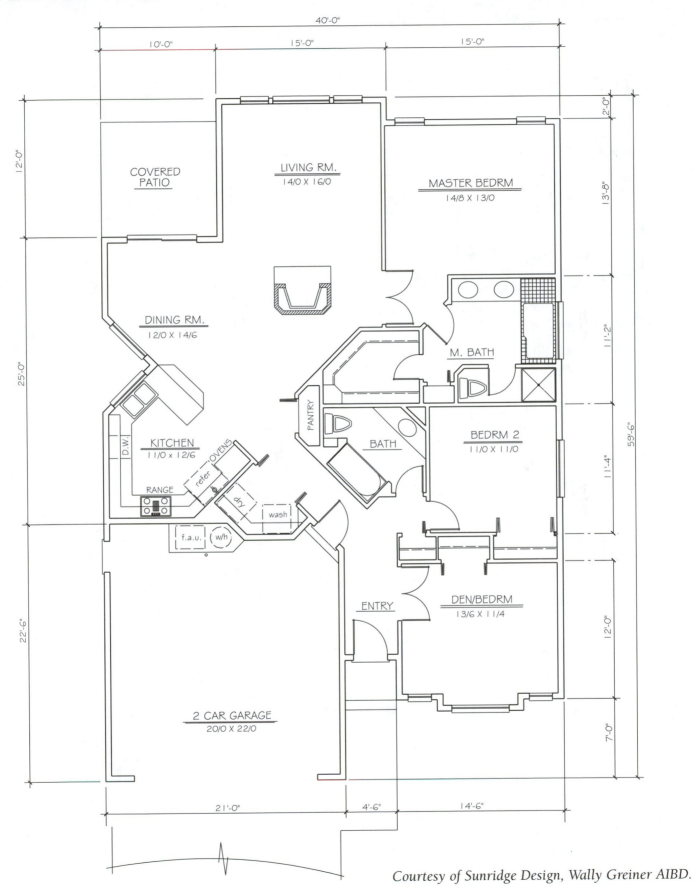

Courtesy of Sunridge Design, Wally Greiner AIBD.

PROBLEM 4-13

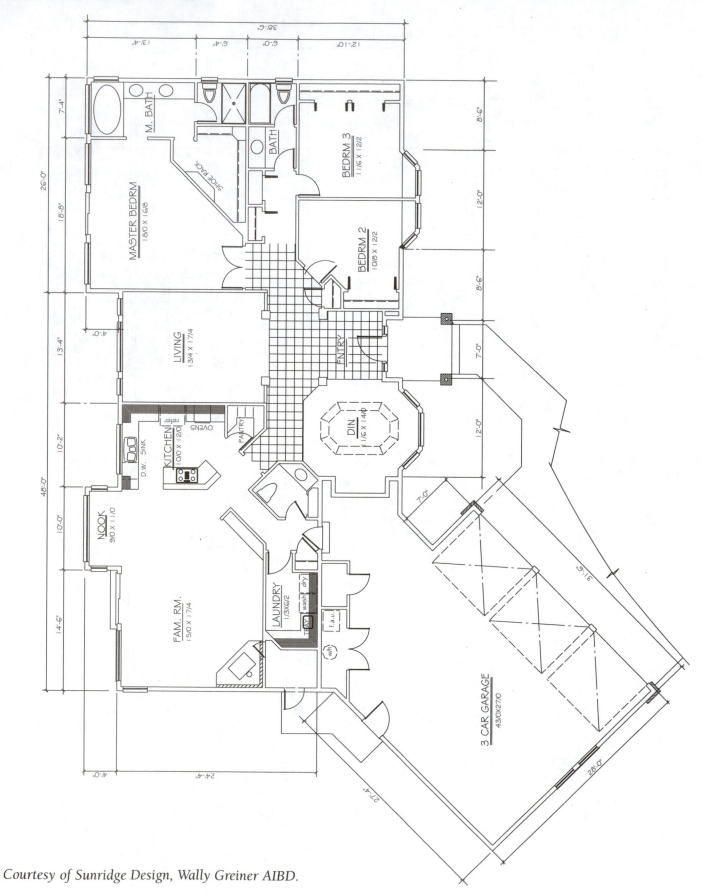

Courtesy of Sunridge Design, Wally Greiner AIBD.

PROBLEM 4-15(B)

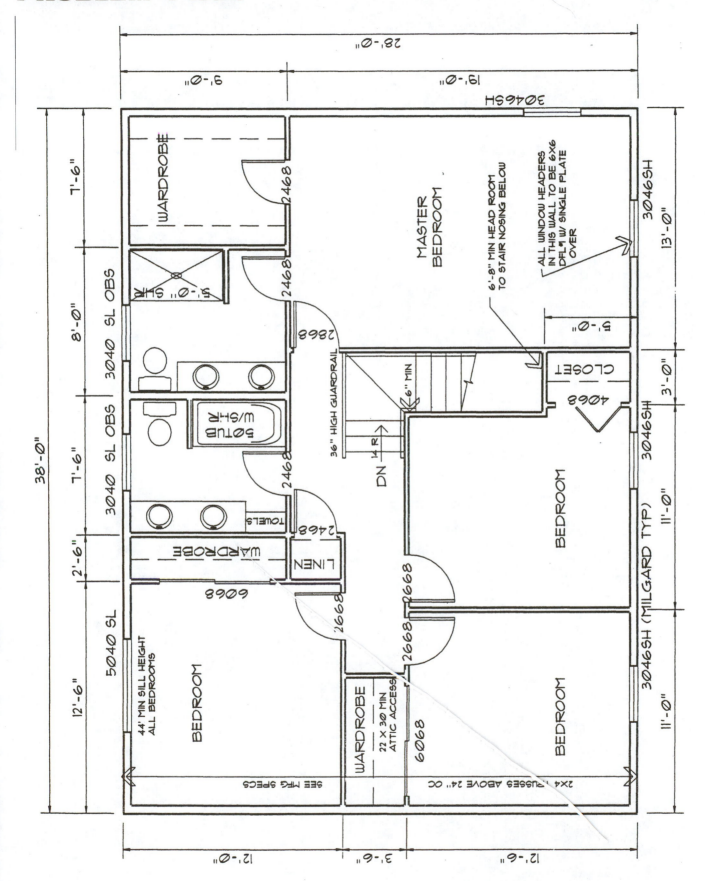

CHAPTER 5

Floor Plan Dimensions

PROBLEM 5-1

Given the three partial floor plans below, provide the exterior dimensions with the dimension-line terminators specified for each. Measure the plans directly using a scale of 1/4" = 1'-0" to establish the dimension numerals.

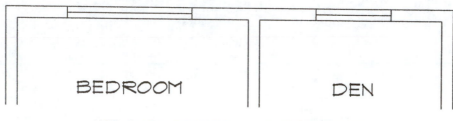

USE SLASH DIMENSION LINE TERMINATORS

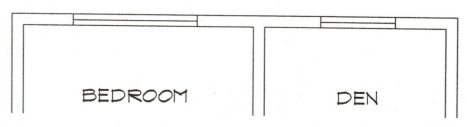

USE ARROWHEAD DIMENSION LINE TERMINATORS

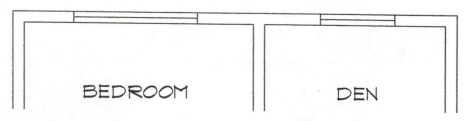

USE DOT DIMENSION LINE TERMINATORS

PROBLEM 5-2

Given the partial floor plan below, provide the necessary exterior and interior dimensions. Use dimension line terminators as specified by your instructor. Measure the drawing directly, using a scale of 1/4" = 1'-0" to establish the dimension numerals.

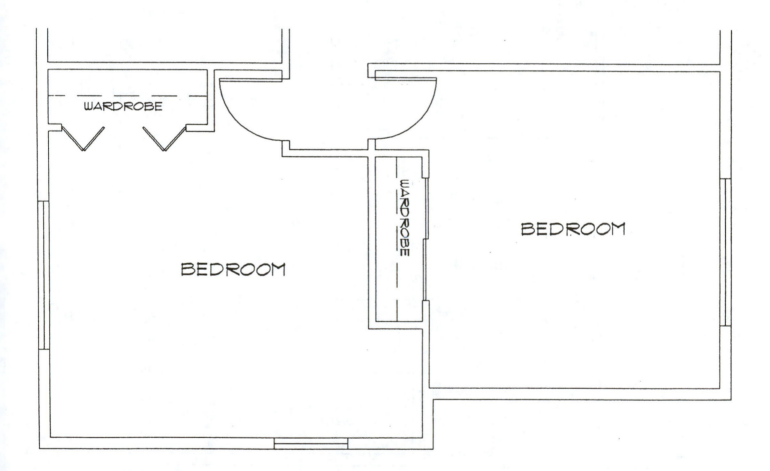

PROBLEM 5-3

Given the partial floor plan below, provide the necessary exterior and interior dimensions. Measure the drawing directly using a scale of 1/4" = 1'-0" to establish the dimension numerals.

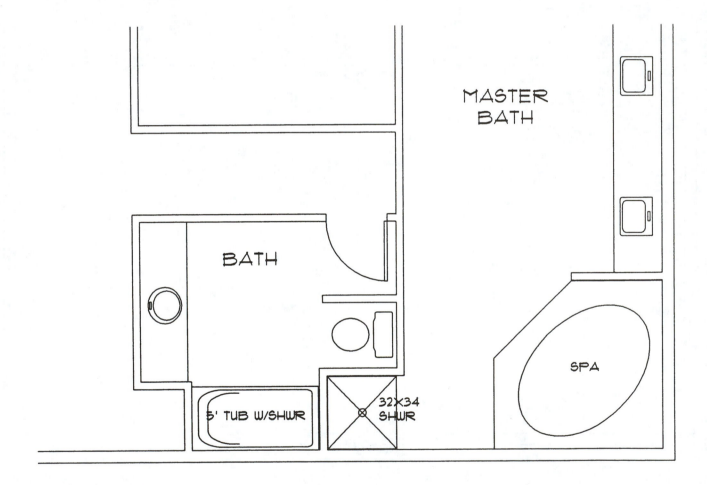

MASTER
BATH

BATH

SPA

5' TUB W/SHWR

32X34
SHWR

PROBLEM 5-4

Given the partial floor plan below, provide the necessary exterior and interior dimensions. Label the brick veneer located in three places on the front of the house. Measure the drawing directly, using a scale of 1/4" = 1'-0" to establish the dimension numerals.

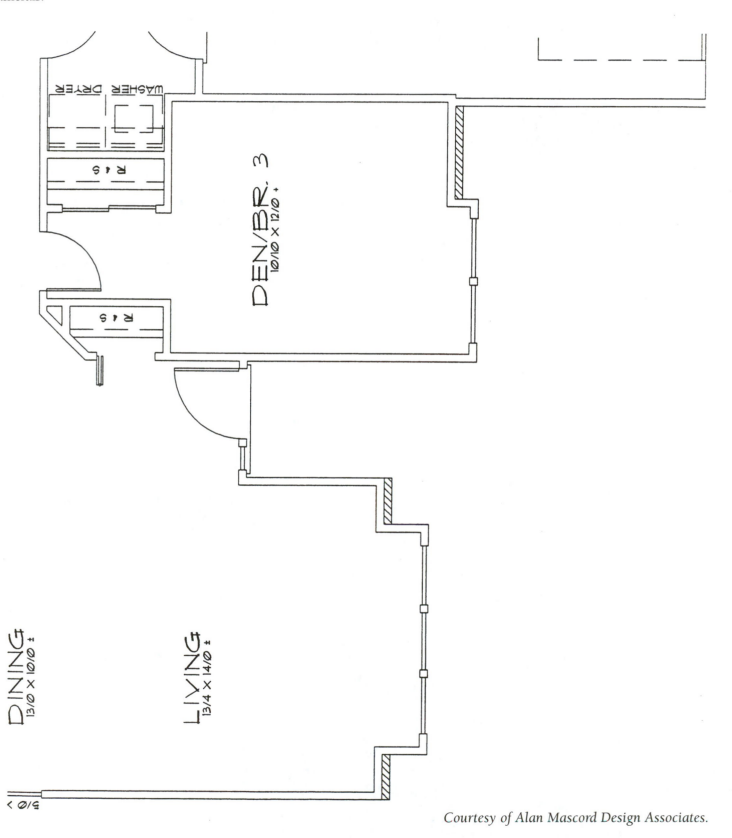

Courtesy of Alan Mascord Design Associates.

PROBLEM 5-5

Given the partial floor plan below using concrete-block construction for the exterior walls, provide the exterior and interior dimensions. Measure the drawing directly using a scale of 1/4" = 1'0" to establish the dimension numerals.

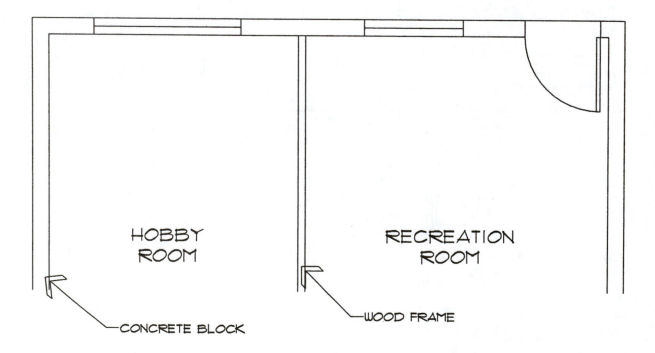

HOBBY
ROOM

RECREATION
ROOM

CONCRETE BLOCK

WOOD FRAME

PROBLEM 5-6

Use the attached partial floor plan or the floor plan from the web site and provide the required interior and exterior dimensions. Assign room names and sizes where appropriate. Represent appliances, plumbing fixtures, and text suitable for a framing plan. Provide a bay window, a pair of 36" wide windows separated by a 4 x 4 post, or a 72" wide slider in each habitable room. Because the printed floor plan is not printed to scale, use X'-X" to represent actual dimensions.

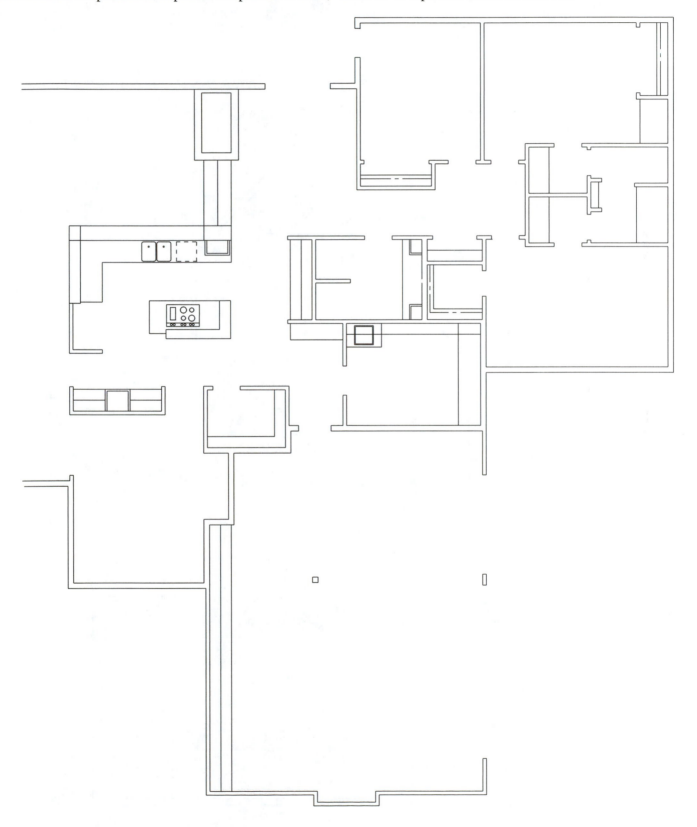

PROBLEM 5-7

Use the attached floor plan or the floor plan from the web site and provide the required interior and exterior dimensions. Assign room names and sizes where appropriate. Represent all required materials and text suitable for a framing plan. Because the printed floor plan is not printed to scale, use X'-X" to represent actual dimensions.

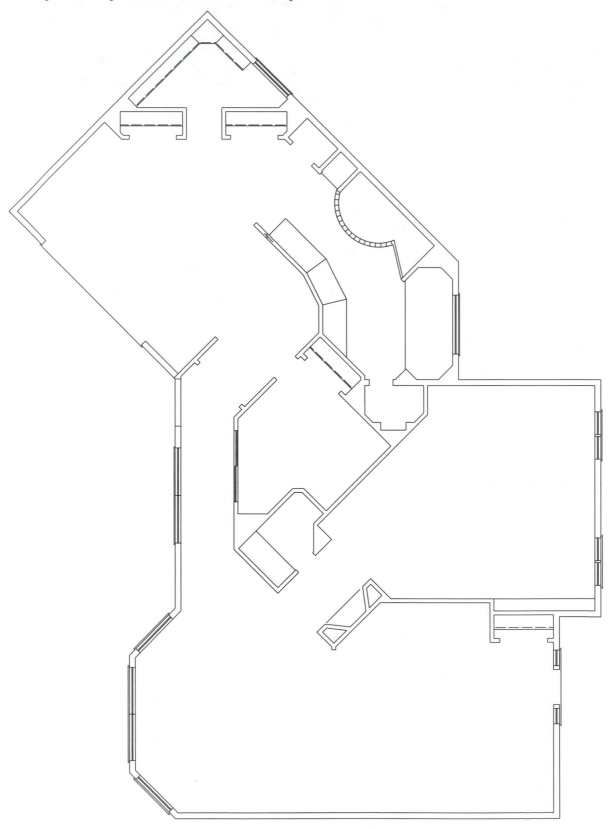

CHAPTER 6

Electrical Plans

PROBLEM 6-1

Name each electrical symbol in the spaces provided below. Use professional style architectural lettering.

_____

_____

_____

_____

_____

_____

_____

_____

_____

_____

_____

_____

_____

_____

_____

PROBLEM 6-2

In the spaces provided below draw a floor-plan representation of the typical electrical installations specified.

PROBLEM 6-3

In the spaces provided below draw a floor-plan representation of the typical electrical installations specified.

SINGLE-POLE SWITCH CONNECTED TO TWO CEILING LIGHTS.

FOUR-WAY SWITCHES CONTROLLING ONE CEILING LIGHT.

THREE-WAY SWITCHES CONNECTED TO ONE CEILING LIGHT.

SINGLE-POLE SWITCH TO A WALL MOUNTED LIGHT AT AN ENTRY.

THREE-WAY SWITCHES CONNECTED TO TWO CEILING LIGHTS.

SINGLE-POLE SWITCH TO A SPLIT-WIRED OUTLET.

PROBLEM 6-4

Access the electrical symbols from the web site and convert each symbol from a drawing to a wblock. Adjust the text size to be suitable for plotting at a scale of 1/4" = 1'-0". Provide each symbol with an appropriate insertion point. Save each wblock using a name that will describe the file.

A. ϕ

B. ϕ^4

C. ϕ^{220}

D. ϕ

E. ϕ^{WP}

F. ϕ^{GFI}

G. ⊙

H. Ⓙ

J. O

K. ⊕

L. ⊙

M. ▢

N. ⊷

P. (rectangle with X and circle)

Q. (square with X and circle)

R. ⊨⎓

S. (symbol with T)

T. $

U. ^{3}

V. ^{D}

W. ▽

X. ∇^I

Y. Ⓟ

Z. Ⓢ

AA. ● S.D.

BB. (circles L H F)

PROBLEM 6-5

Given the partial floor plan on page 80, provide the electrical layout as follows:

- Provide at least six duplex convenience outlets in the master bedroom. Place two of the six split-wired outlets along the long left wall, and control these two outlets with three-way switches, one switch located as you enter the room and the other between the bedroom and the master bath.

- Place a light over each master bath sink, controlled by a single switch.

- Place a duplex convenience outlet by each vanity sink.

- Provide an exhaust fan in the master bath shower/water closet compartment.

- Place a recessed GFCI ceiling light in the master bath shower/water closet compartment.

- Provide a ceiling light with a switch for the master walk-in wardrobe.

- Center two recessed GFCI ceiling lights over the spa with a switch conveniently located.

- Provide an exhaust fan/heat lamp and a control switch in the family bath.

- Provide a wall-mounted light over the sink in the family bath with a conveniently located switch.

- Provide two ceiling lights in the hallway to the bedrooms with three-way switches conveniently located.

- Provide a ceiling light in bedroom 2 with a switch next to the door.

- Place at least two duplex convenience outlets in bedroom 2 on the walls shown.

- Provide the needed electrical for the utility room.

PROBLEM 6-5 (CONTINUED)

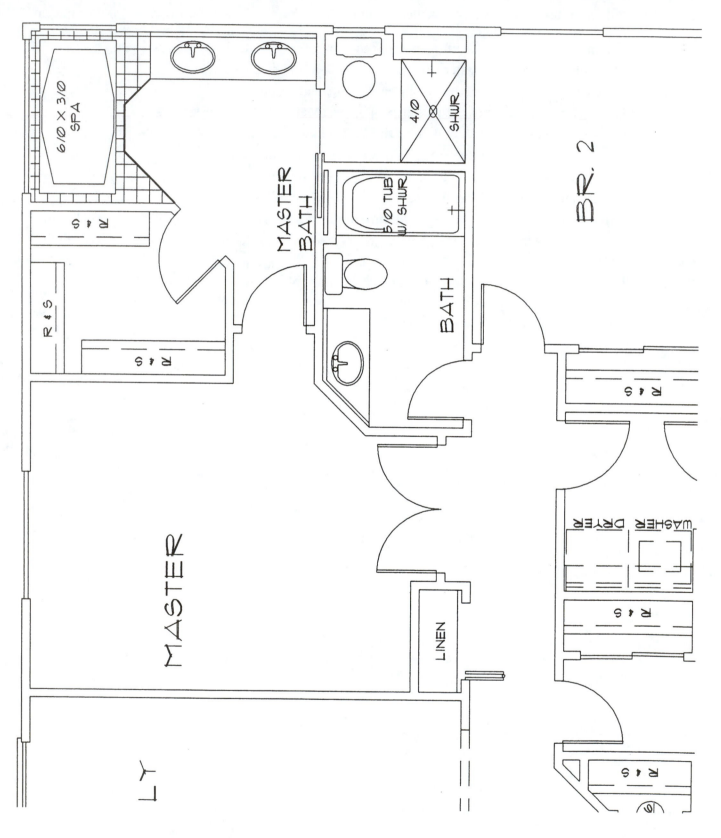

Courtesy of Alan Mascord Design Associates.

PROBLEM 6-6

Given the partial floor plan on page 82, provide the electrical layout as follows:

- Provide switches to control the following fixtures:
- Center a ceiling light in the foyer with a switch in the foyer below and another switch in a convenient location at the top of the stairs.
- Place at least three equally spaced recessed ceiling lights in the balcony/hallway area with one switch by the master bedroom door, a switch near bedroom 4 door, and a third switch near bedroom 2 door.
- Place a wall-mounted light above each vanity sink.
- Provide a duplex convenience outlet by each vanity sink.
- Place two recessed ceiling lights centered above the spa.
- Provide a ceiling light/fan combination in the area between the shower and water closet.
- Provide a ceiling light in the master wardrobe.
- Provide a ceiling light centered in bedrooms 2, 3, and 4.
- Locate duplex convenience outlets in bedrooms 1, 2, 3 and 4.
- Place at least two duplex convenience outlets in the balcony/hallway.
- Place ceiling-mounted smoke detectors near the top of the stairs and for each bedroom.
- Provide an exhaust fan/heat light combination with a switch in the family bath.

PROBLEM 6-6 (CONTINUED)

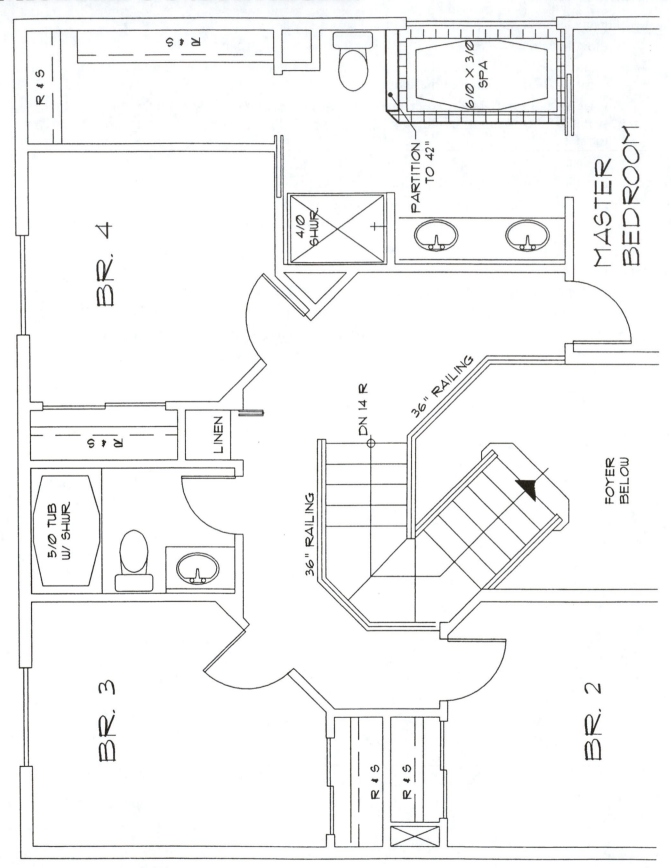

Courtesy of Alan Mascord Design Associates.

PROBLEM 6-7

Given the partial floor plan on page 84, provide the electrical layout as follows:

- 220v outlet at range.
- Dishwasher outlet.
- Refrigerator outlet.
- Recessed ceiling light above sink with single-pole switch.
- Four recessed ceiling lights placed to provide uniform lighting in the kitchen with single-pole switch at short wall next to pantry.
- Fluorescent fixtures mounted under the upper cabinets.
- Duplex convenience outlet in island front.
- Ceiling light centered in nook with three-way switch, one placed inside patio door at family room and the other at short wall next to pantry.
- Four duplex convenience outlets above kitchen base cabinets.
- Two duplex convenience outlets on right wall of family room.
- Two duplex convenience outlets, one on each side of the fireplace.
- Three duplex convenience outlets in the dining room.
- A recessed ceiling light centered above and in front of the fireplace with a single-pole switch next to the fireplace.
- An outside wall-mounted light next to the patio door in the family room with a single-pole switch on the inside.
- Waterproof duplex convenience outlet outside the patio door in the family room.

PROBLEM 6-7 (CONTINUED)

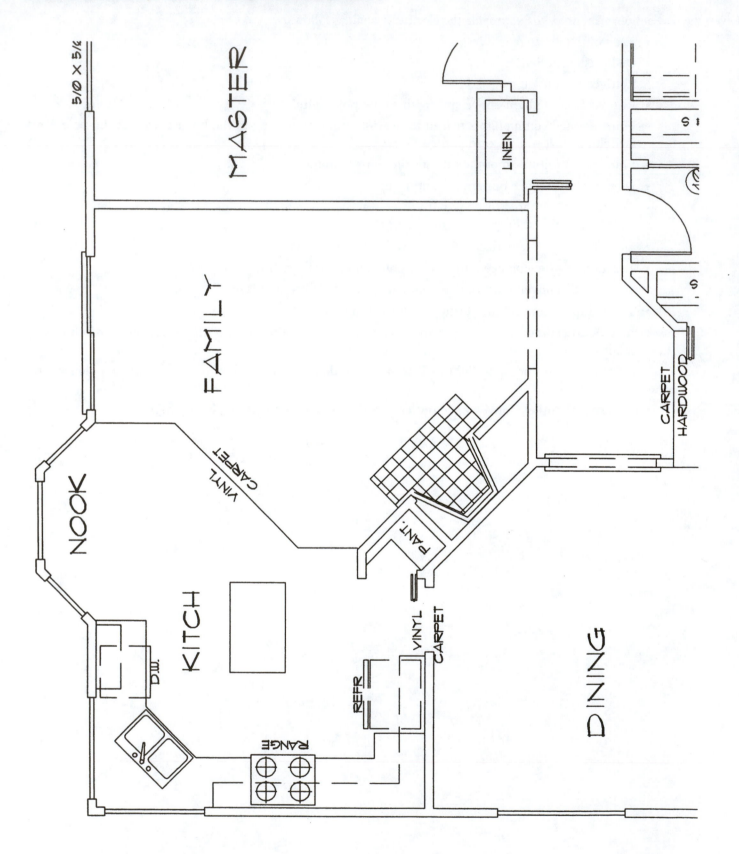

Courtesy of Alan Mascord Design Associates.

PROBLEM 6-8

Access the floor plan for Problem 5-6 from the web site. Create new layers for electrical symbols, wiring, and notes. Attach the electrical notes from the web site to the drawing and edit the notes as required to meet the needs of your drawing. Use the blocks created in Problem 6-4 and complete the electrical plan. Eliminate all symbols and notes that do not apply to your drawing. Set all variables for plotting at a scale of 1/4" = 1'–0". Save each wblock using a name that will describe the file.

PROBLEM 6-9

Access the floor plan for Problem 5-7 from the web site. Create new layers for electrical symbols, wiring, and notes. Attach the electrical notes from the web site for problem 6-8 to the drawing and edit the notes as required to meet the needs of your drawing. Use the blocks created in Problem 6-4 and complete the electrical plan. Eliminate all symbols and notes that do not apply to your drawing. Set all variables for plotting at a scale of 1/4" = 1'–0". Save each wblock using a name that will describe the file.

PROBLEM 6-10

Use the floor plan from the web site that was started in Chapter 4 to complete an electrical plan. Freeze all layers such as text or dimensions that were required for the floor plan. Create new layers for electrical symbols, wiring, and notes. Attach the electrical notes and legend from the web site for problem 6-8 to the drawing and edit the notes as required to meet the needs of your drawing. Use the blocks created in Problem 6-4 to complete the electrical plan. Eliminate all symbols and notes from the legend that do not apply to your drawing. Set all variables for plotting at a scale of 1/4" = 1'–0".

CHAPTER 7
Plumbing and HVAC Drawings

PROBLEM 7-1

Define the following plumbing terms.

CW	HW	HWR
HB	CO	DS

PROBLEM 7-2

Define the following plumbing fixtures.

CB	MH	VTR
DF	WH	DW
BD	GD	
WC	LAV	B
S	U	SH
RD	FD	SD

PROBLEM 7-3

Use the floor plan that was started in Chapter 4 and draw a one-line isometric drawing of the fresh water supply system.

PROBLEM 7-4

Use the symbols in Figure 18-4 from your text and create a wblock of each of the symbols. Assign an insertion point appropriate for easy insertion into a plumbing layout. Create the symbols at a size appropriate for inserting in a drawing plotted at a scale of 1/4" = 1'-0". Save each symbol with a file name that will describe the symbol.

PROBLEM 7-5

Use the floor plan that was started in Chapter 4 and draw a one-line isometric drawing of the drainage and waste system.

PROBLEM 7-6

Use vender supply catalogs or the Internet and complete a plumbing fixture schedule for the residence that was started in Chapter 4.

PROBLEM 7-7

Use the symbols in Figure 19-13 from your text and create a wblock of each of the symbols. Assign an insertion point appropriate for easy insertion into a HVAC layout. Create the symbols at a size appropriate for inserting in a drawing plotted at a scale of 1/4" = 1'-0". Save each symbol with a file name that will describe the symbol.

PROBLEM 7-8

Use the symbols in Figure 19-24 from your text and create a wblock that can be used for an exhaust grill schedule. Assign an insertion point appropriate for easy insertion into a HVAC layout. Create the grids and text at a size appropriate for inserting in a drawing plotted at a scale of 1/4" = 1'-0". Save each schedule with an appropriate file name

PROBLEM 7-9

Use the floor plan that was started in Chapter 4 to design and draw a mechanical plan. Freeze all layers that were required for the floor and electrical plans. Create new layers for HVAC symbols, and notes. Use the blocks created in Problem 7-7 to complete the plan. Set all variables for plotting at a scale of 1/4" = 1'–0".   Base all design assumptions on the outdoor temperature for your area and assume the minimum code allowed indoor temperature. Ceiling heights are to be based on your design. Insulation, doors, weather stripping, and glazing to meet or exceed minimum code requirements for your area.

PROBLEM 7-10

Use the floor plan that was started in Chapter 4 and prepare the thermal calculations for heating and cooling using the blank forms that are provided on the web site for Architectural Drafting & Design. Use the guidelines for problem 7-8 to complete this project.

PROBLEM 8-7

Draw a roof plan showing a hip roof with 24" overhangs.

PROBLEM 8-8

Draw a roof plan showing a Dutch hip roof with 24" overhangs. Use 12" overhangs at the gable-end walls. Use guide lines for eave placement.

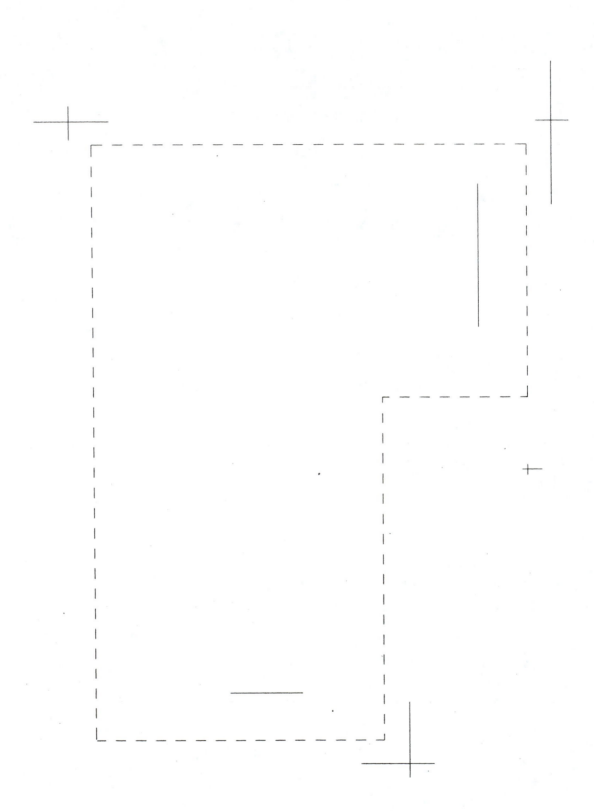

PROBLEM 8-9

Draw a roof plan showing a hip roof with 24" overhangs. Show all required vents and downspouts. Draw a chimney somewhere near a ridge.

PROBLEM 8-10

Draw a roof plan showing a hip roof with 24" overhangs.

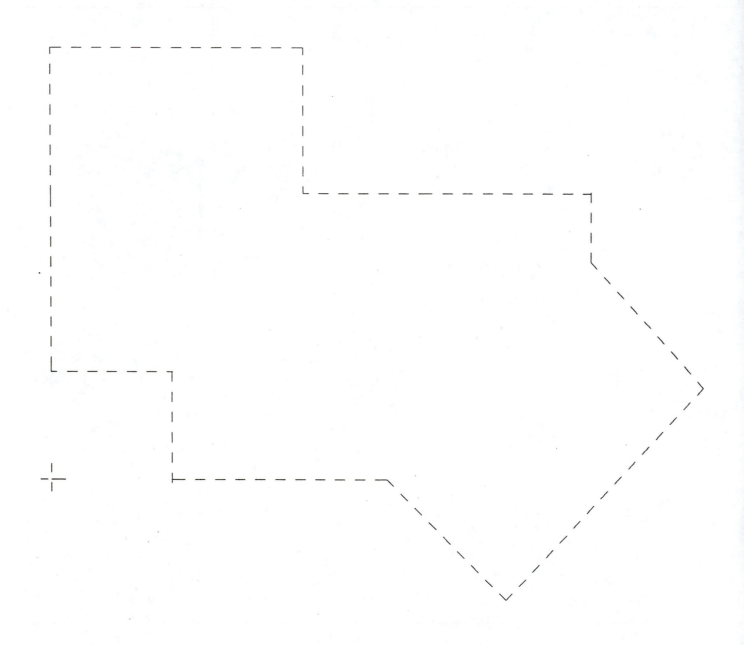

PROBLEM 8-11

Use a scale of 1/16" = 1'0" to complete the following roof plans. Use the drawing below and design three different roofs to cover this residence. Draw all items typical for the type of plan and label all material.

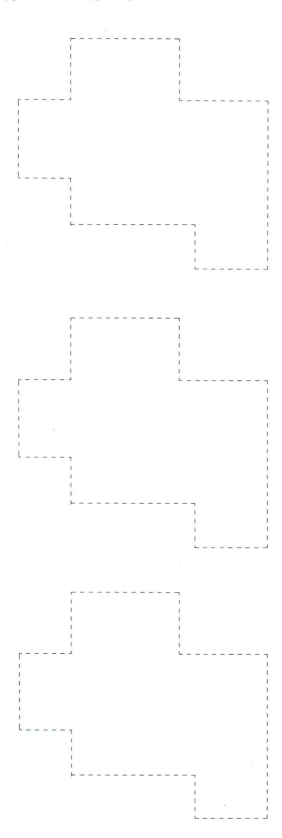

PROBLEM 8-12

Draw a roof plan showing two Dutch Hips with 24" and 12" overhangs where appropriate. Show rafter spans and headers where required. Show a 4' x 4' skylight, a chimney, roof vents, and downspouts.

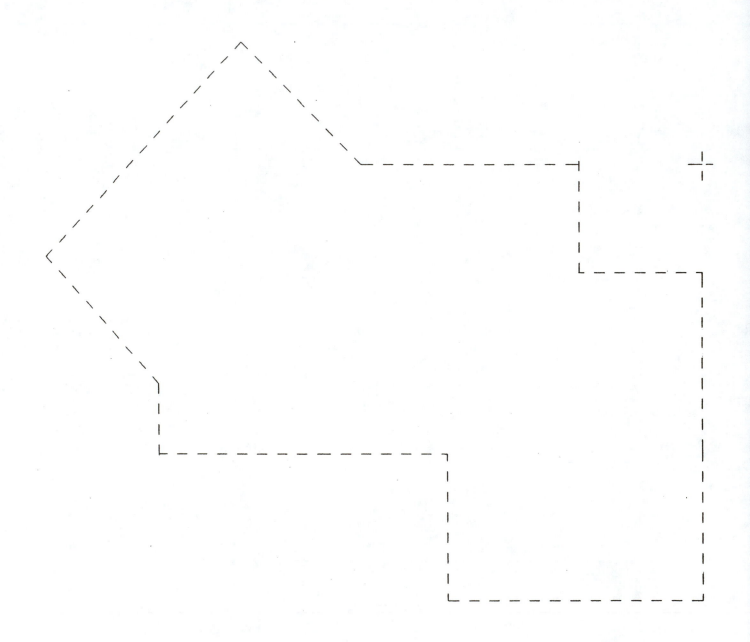

PROBLEM 8-13

Download the roof notes from the Web site and edit them to meet the needs of your project. Attach the edited notes of the following Roof Plans.

ROOF PLAN

1/8" ———— 1'-0"

ALL FRAMING LUMBER TO BE DFL #2 OR
BETTER UNLESS NOTED.

ALL RAFTERS TO BE 2 X 6 UNLESS NOTED.
SEE ATTACHED SCHEDULE FOR SPECIFIC
SIZES

SUBMIT TRUSS MANUF. DRAWINGS TO BUILDING
DEPT. PRIOR TO ERECTION

PROVIDE SCREENED VENTS @ EA. 3rd.
JOIST SPACE @ ALL ATTIC EAVES.

USE 1/2" 'CCX' EXTERIOR PLY @ ALL
EXPOSED EAVES

USE 235# COMPO. SHINGLES OVER 15# FELT.

USE 300# COMPO. SHINGLES OVER 15# FELT.

USE MED. CEDAR SHAKES OVER 15# FELT W/
30# x 18" WIDE FELT BTWN. EA. COURSE. W/
10 1/2" EXPOSURE.

USE MONIER HOMESTEAD NATURAL CHARCOAL
ROOF TILES OVER 15 # FELT. INSTALL AS PER
MANUF SPECS. VERIFY COLOR AND STYLE W/ OWNER

DIRECTIONS: For Problems 8-14 through 8-16:

1. Using prints of your floor plans, complete the following problems using the drawings as a guide only. If your elevations and sections are already drawn, be sure that the overhangs match on all drawings. All roofs will be framed with trusses unless noted.

2. Use a scale suitable for the required plan. If your plan can be drawn at a scale of 1/8" = 1'-0", place the drawing on the vellum so that another drawing can be placed on the same sheet.

3. When you have evaluated your drawing, hand in a print of it to your instructor for evaluation.

PROBLEM 8-14(A)

Draw the location of trusses on the floor plan drawn in Problem 4-1 (page 50).

PROBLEM 8-14(B)

Draw a roof plan for the residence drawn in Problem 4-1.

PROBLEM 8-14(C)

Draw a roof framing plan for the residence drawn in Problem 4-1.

PROBLEM 8-15(A)

Draw the location of trusses on the floor plan drawn in Problem 4-2 (page 51).

PROBLEM 8-15(B)

Draw a roof plan for the residence drawn in Problem 4-2.

PROBLEM 8-15(C)

Draw a roof framing plan for the residence drawn in Problem 4-2.

PROBLEM 8-16(A)

Draw the location of trusses on the floor plan drawn in Problem 4-3 (page 52).

PROBLEM 8-16(B)

Draw a roof plan for the residence drawn in Problem 4-3.

PROBLEM 8-16(C)

Draw a roof framing plan for the residence drawn in Problem 4-3.

PROBLEM 8-17(A)

Draw the location of rafters on the floor plan drawn in Problem 4-4 (page 53).

PROBLEM 8-17(B)

Draw a roof framing plan for the residence drawn in Problem 4-4.

PROBLEM 8-18(A)

Draw the location of rafters on the floor plan drawn in Problem 4-5 (page 54).

PROBLEM 8-18(B)

Draw a roof plan for the residence drawn in Problem 4-5.

PROBLEM 8-19(A)

Draw the location of trusses on the floor plan drawn in Problem 4-6 (page 55).

PROBLEM 8-19(B)

Draw a roof plan for the residence drawn in Problem 4-6.

PROBLEM 8-20(A)

Draw the location of ceiling joists on the floor plan drawn in Problem 4-7 (page 56).

PROBLEM 8-20(B)

Draw a roof framing plan for the residence drawn in Problem 4-7.

PROBLEM 8-21(A)

Draw the location of trusses on the floor plan drawn in Problem 4-8 (page 57).

PROBLEM 8-21(B)

Draw a roof framing plan for the residence drawn in Problem 4-8.

PROBLEM 8-22(A)

Draw the location of rafters on the floor plan drawn in Problem 4-9 (page 58).

PROBLEM 8-22(B)

Draw a roof framing plan for the residence drawn in Problem 4-9.

DIRECTIONS for Problems 8-23 through 8-27:
Use the floor plan that was started in Chapter 4 as a base to draw a roof plan. Use the preliminary exterior elevation in Chapter 9 as a guide to determine roof shapes. The roof design can be altered from the shape presented in Chapter 9 if the elevations are altered to match the floor and roof plans. Freeze all material on the floor and electrical plans and use the base floor plan as a starting point for creating the roof plan. Create new layers on the floor plan drawing for the roof material. Set all text and line scale values for plotting at a scale of 1/8" = 1'–0". Insert the roof notes from the web site, and edit them to meet the needs of your project.

DIRECTIONS for Problems 8-28 through 8-32:
Use the roof plan that was created in problems 8-23-through 8-27 as a base to complete the roof framing plan. Create new layers for framing members. Set all text and line scale values for plotting at a scale of 1/8" = 1'–0". Insert and edit the rafter schedule from the web site as needed.

PROBLEM 8-33

Draw a roof plan for the residence drawn in Problem 4-15 (page 64).

PROBLEM 8-34

Download the rafter schedule from the web site, and edit the schedule to meet the needs of your project. Attach the schedule to your roof framing plan.

DESIGN STANDARDS

BASED ON 2000 IBC TABLE R802.5.1(1)
Douglas fir-larch #2

RAFTERS: TABLE 7-O

10# DEAD LOAD / 20 # LIVE LOAD
2 x 6 @ 12" O.C. = 16'-7" MAX.
2 x 6 @ 16" O.C. = 14'-4" MAX.
2 x 6 @ 24" O.C. = 11'-9" MAX.

2 x 8 @ 12" O.C. = 21'-0" MAX.
2 x 8 @ 16" O.C. = 18'-2" MAX.
2 x 8 @ 24" O.C. = 14-'10" MAX

2 x 10 @ 12" O.C. = 25'-8" MAX.
2 x 10 @ 16" O.C. = 22'-3" MAX.
2 x 10 @ 24" O.C. = 18-2" MAX.

2 x 12 @ 16" O.C. = 25'-9" MAX.
2 x 12 @ 24" O.C. = 18-'2" MAX

CHAPTER 9

Elevation Fundamentals

DIRECTIONS: Complete the following problems on a separate sheet of vellum. Assume 1/4" = 1'-0" for all drawing problems unless otherwise noted. Use a proper style of architectural lettering to specify any specified materials or dimensions.

PROBLEM 9-1

Draw an example of a wall covered with T1-11, with grooves at 8" o.c., and show a step in the foundation of 14" and a sloping grade.

PROBLEM 9-2

Draw a 6' x 4' sliding window with 15"-wide shutters on each side. Surround with 6" horizontal siding.

PROBLEM 9-3

Draw a 3'-0" x 6'-8" door with an oval glass insert, with 12" x 48" wide side lites.

PROBLEM 9-4

Draw a 3'-0" x 6'-8" window with the lower 2'-0" awning and a half-round transom above it.

PROBLEM 9-5

Draw a 3-'0" x 3'-6" double-hung window with grids and 15"-wide shutters.

PROBLEM 9-6

Draw an example of a gable-end wall, a 5/12 pitch and a metal chimney in a 4' x 2' chase with horizontal wood siding. Properly dimension the chimney and ridge.

PROBLEM 9-7

Draw an example of the bottom of a wall covered with 6"-wide bevel siding over Tyvek.. Use 1x3 corner trim and show the line of the finish floor (assume a 2 x 8 floor joist), and show the bottoms of footings (assume footings extend 18" into grade for frost line).

PROBLEM 9-8

Draw clay tiles on the gable end of a roof with a 6/12 pitch. Specify a 2 x 6 barge rafter and 1" exterior stucco.

PROBLEM 9-9

Draw a 36"-high rail made from 2 x 2 verticals with a 2 x 6 wood top rail.

PROBLEM 9-10

Draw an 18"-square column with 1" exterior stucco. Design a built-up top and bottom. Use a scale of 1/2" = 1'-0".

PROBLEM 9-11

A client would like to have a 6' x 4' picture window with 15"-wide shutters on each side. The ceiling height is 8'-0". The roof pitch is 7/12. The owner would like to have a half-round transom window above the double-hung windows. Design an elevation showing a dormer gable roof above these windows perpendicular to the main roof. Draw the dormer plate height as 10". Surround with 6" horizontal siding.

PROBLEM 9-12

Draw a pair of 3'-0" x 6'-8" windows with the lower 1'-6" awning. Surround with 1 x 3 trim with wood shingles above to a 2 x 8 fascia. Show 1" exterior stucco on each side.

PROBLEM 9-13

Draw an example of a post and 36"-high rail that could be used on a porch of a Victorian style residence. Show decorative trim at the top of the post.

PROBLEM 9-14

Draw an example of a 6'-0" x 6'-8" 10-lite French door opening at a deck. Design and show a solid railing that would blend with an English Tudor style home.

PROBLEM 9-15

Draw an example of a gable end wall, a 2 x 6 barge rafter, 4/12 pitch, and medium cedar shakes.

DIRECTIONS: Problems 9–16 through 9–18 are typical of design changes required when working with a stock plan that will be built several times in a subdivision. Use the drawing as a base and design a front elevation to meet the given criteria. Working with a base elevation, design an exterior to match the following criteria. Complete Problems 9–16 and 9–17 without making changes in the design. Only exterior finishing materials may be altered.

PROBLEM 9-16(A)

Complete the elevation using horizontal siding and trim to look traditional.

PROBLEM 9-16(B)

Complete the elevation using materials to give the house a Spanish look.

PROBLEM 9-16(C)

Complete the elevation using materials to give the house a contemporary style.

PROBLEM 9-22

Using your floor plan (see Problem 4-2) and the drawings below, draw the required elevations. Pick siding and roofing appropriate for your area. Draw the front elevation as a presentation elevation.

PROBLEM 9-23

Using your floor plan (see Problem 4-3) and the drawings below, draw the required elevations. Pick siding and roofing appropriate for your area. Draw the front elevation as a presentation elevation and the other elevations as working elevations.

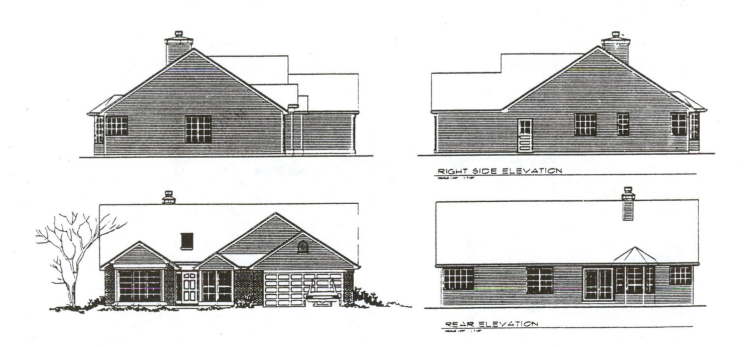

PROBLEM 9-24

Using your floor plan (see Problem 4-4) and the drawings below, draw the required elevations. Pick siding and roofing appropriate for your area. Draw the front elevation as a presentation elevation.

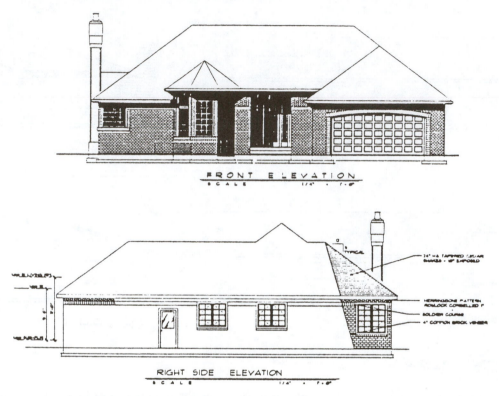

PROBLEM 9-25

Using your floor plan (see Problem 4-5) and the drawings below, draw the required elevations. Pick siding and roofing appropriate for your area. Draw the front elevation as a presentation elevation.

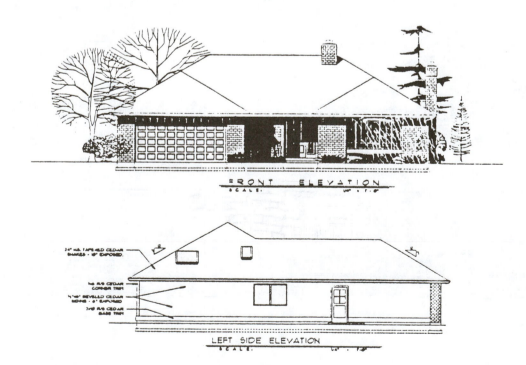

PROBLEM 9-26

Using your floor plan (see problem 4-6) and the drawings below, draw the required elevations. Pick siding and roofing appropriate for your area. Draw the front elevation as a presentation elevation.

FRONT ELEVATION
SCALE: 1/4"=1'-0"

RIGHT ELEVATION
SCALE: 1/4"=1'-0"

PROBLEM 9-27

Using your floor plan (see Problem 4-7) and the drawing below, draw the required elevations. Pick siding and roofing appropriate for your area. Draw the front elevation as a presentation elevation.

FRONT ELEVATION
SCALE   1/4" = 1'-0"

PROBLEM 9-28

Using your floor plan (see Problem 4-8) and the sketch below, draw the required elevations. Pick siding and roofing appropriate for your area. Draw the front elevation as a presentation elevation.

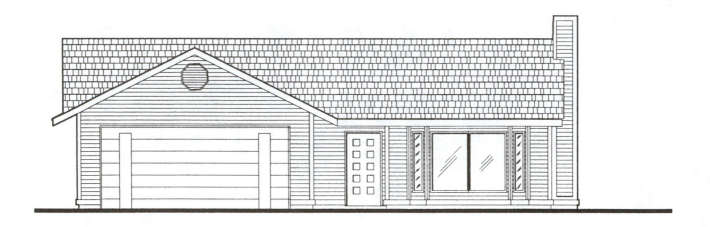

FRONT ELEVATION

scale 1/4" = 1'-0"

PROBLEM 9-29

Using your floor plan (see Problem 4-9) and the sketch below, draw the required elevations. Pick siding and roofing appropriate for your area. Draw the front elevation as a presentation elevation.

FRONT ELEVATION

PROBLEM 9-30

Using your floor plan (see Problem 4-10) and the sketch below, draw the required elevations. Pick siding and roofing appropriate for your area. Draw the front elevation as a presentation elevation.

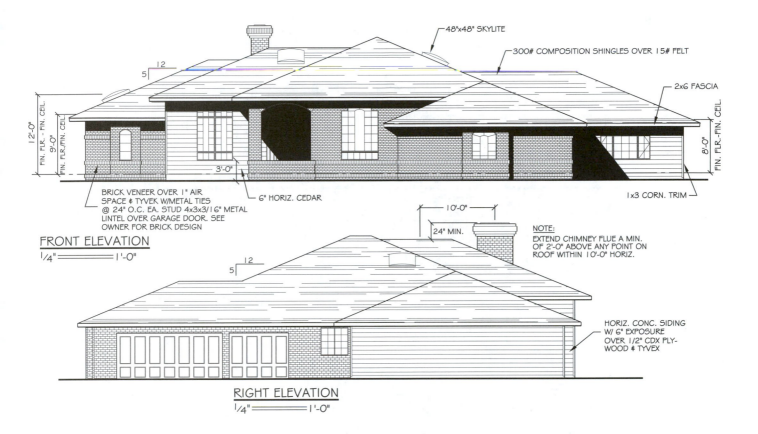

48"x48" SKYLITE

300# COMPOSITION SHINGLES OVER 15# FELT

2x6 FASCIA

FIN. FLR. - FIN. CEIL. 12'-0"

FIN. FLR./FIN. CEIL. 9'-0"

FIN. FLR.-FIN. CEIL. 8'-0"

12 / 5

3'-0"

6" HORIZ. CEDAR

BRICK VENEER OVER 1" AIR SPACE & TYVEK W/METAL TIES @ 24" O.C. EA. STUD 4x3x3/16" METAL LINTEL OVER GARAGE DOOR. SEE OWNER FOR BRICK DESIGN

1x3 CORN. TRIM

FRONT ELEVATION
1/4" ══════ 1'-0"

10'-0"

24" MIN.

12 / 5

NOTE:
EXTEND CHIMNEY FLUE A MIN. OF 2'-0" ABOVE ANY POINT ON ROOF WITHIN 10'-0" HORIZ.

HORIZ. CONC. SIDING W/ 6" EXPOSURE OVER 1/2" CDX PLY-WOOD & TYVEX

RIGHT ELEVATION
1/4" ══════ 1'-0"

PROBLEM 9-31

Using your floor plan (see Problem 4-11) and the sketch below, draw the required elevations. Pick siding and roofing appropriate for your area. Draw the front elevation as a presentation elevation.

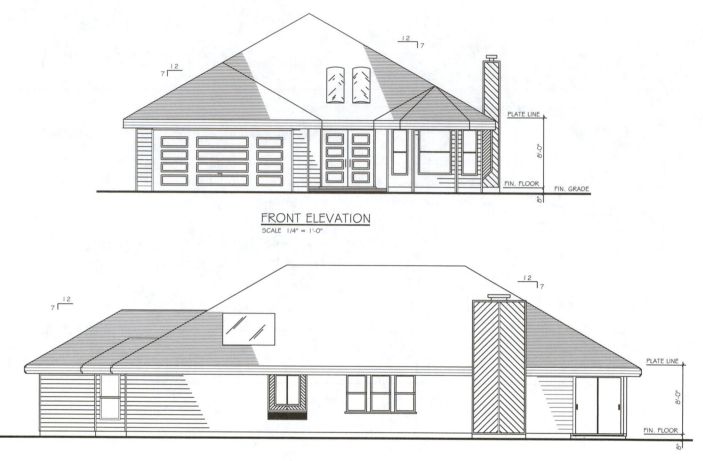

FRONT ELEVATION
SCALE 1/4" = 1'-0"

RIGHT ELEVATION
SCALE 1/4" = 1'-0"

PROBLEM 9-32

Using your floor plan (see Problem 4-12) and the sketch below, draw the required elevations. Pick siding and roofing appropriate for your area. Draw the front elevation as a presentation elevation.

STUCCO SIDING OR EQUAL
OVER 15# A.S. FELT
AT FRONT OF HOUSE

FRONT ELEVATION
SCALE 1/4" = 1'-0"

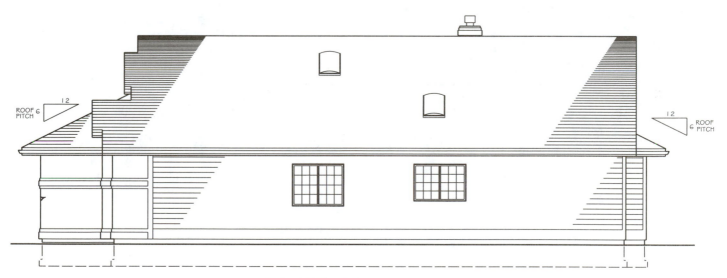

ROOF
PITCH

RIGHT ELEVATION
SCALE 1/4" = 1'-0"

PROBLEM 9-33

Use the floor plan from Problem 4-13, the attached preliminary design, and the guidelines in your text to layout and draw the required elevations to clearly describe the residence. Select siding to match the preliminary design.

FRONT ELEVATION
SCALE 1/4" = 1'-0"

RIGHT ELEVATION
SCALE 1/4" = 1'-0"

PROBLEM 9-34

Use the floor plan from Problem 4-14, the attached preliminary design, and the guidelines in your text to layout and draw an elevation for each plane of the residence. Select siding to match the preliminary design.

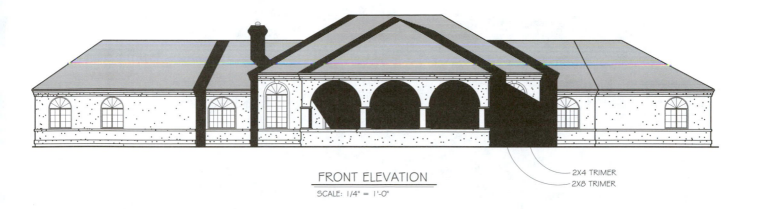

FRONT ELEVATION

SCALE: 1/4" = 1'-0"

2X4 TRIMER
2X8 TRIMER

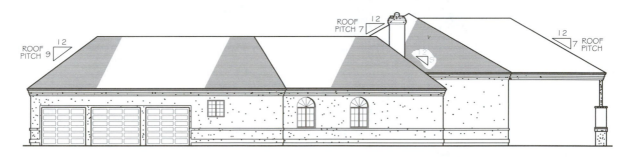

ROOF PITCH 7 12

ROOF PITCH 9 12

12 7 ROOF PITCH

LEFT ELEVATION

SCALE: 1/4" = 1'-0"

PROBLEM 9-35

Using your floor plan (see Problem 4-15) and the guidelines in your text, lay out and draw the required elevations. Pick siding and roofing appropriate for your area.

PROBLEM 10-6

Use the given cabinet elevations as a guide to draw the cabinet elevations for the set of plans that you started in Problem 4-2.

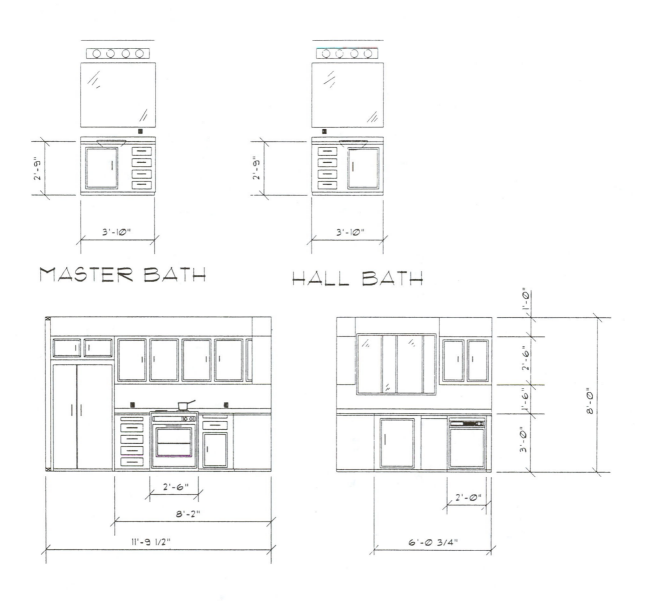

MASTER BATH   HALL BATH

KITCHEN

NOTE: DIMENSIONS GIVEN AS REFERENCE ONLY.
      YOUR ACTUAL DIMENSIONS ARE BASED ON
      YOUR FLOOR PLAN AND MAY BE DIFFERENT.

PROBLEM 10-7

Use the given cabinet elevations as a guide to draw the cabinet elevations for the set of plans that you started in Problem 4-3.

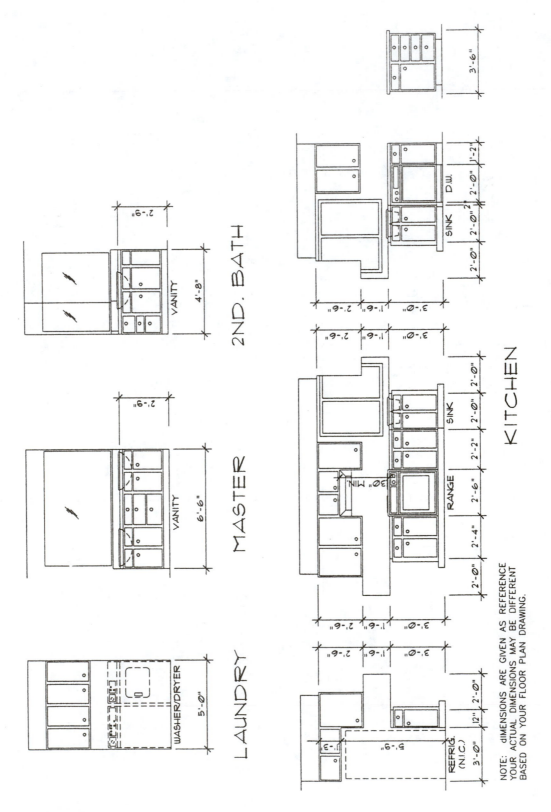

PROBLEM 10-8

Use the given cabinet elevations as a guide to draw the cabinet elevations for the set of plans that you started in Problem 4-4.

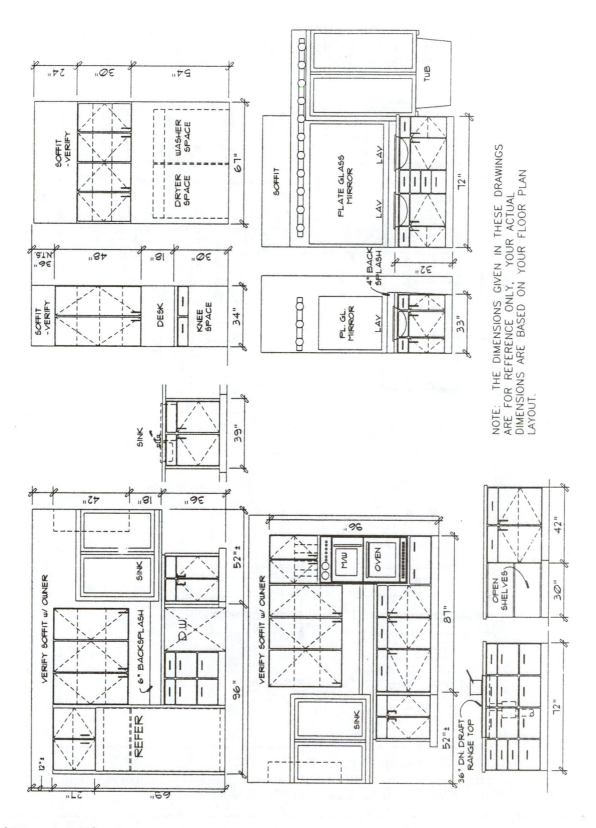

Courtesy of Piercy & Barclay Designers, Inc.

PROBLEM 10-9

Use the given cabinet elevations as a guide to draw the cabinet elevations for the set of plans that you started in Problem 4-5.

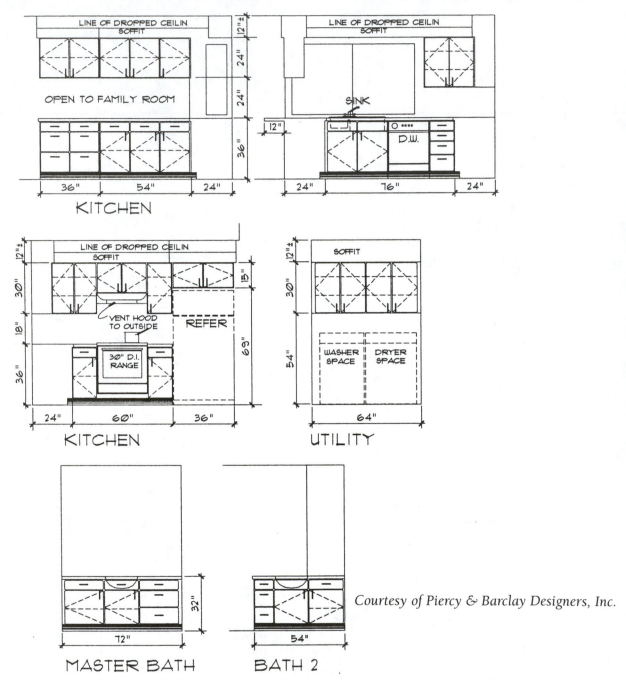

Courtesy of Piercy & Barclay Designers, Inc.

NOTE: DIMENSIONS ARE GIVEN FOR REFERENCE ONLY YOUR ACTUAL DIMENSIONS MAY VARY BASED ON YOUR FLOOR PLAN DRAWING.

Directions Problems 10-10 through 10-15:
Use the floor plan that you drew in Chapter 4 to prepare a complete set of cabinet elevations. The cabinet elevations that you draw should be designed in relation to the floor plan.

CHAPTER 11
Foundation Plans

DIRECTIONS: Use a scale of 1/4" = 1' unless noted otherwise.

PROBLEM 11-1

Using the drawing below as a guide, draw and label the required information to show a plan view of a one-story, continuous footing made of poured concrete supporting a wood floor.

POURED CONCRETE

PROBLEM 11-2

Using the drawing below as a guide, draw and label the required information needed to show a plan view of a one-story, continuous concrete foundation and stem wall made of concrete block.

PROBLEM 11-3

Using the drawing below as a guide, draw and label the required information needed to show a one-story foundation made of poured concrete. Thicken as required to support brick veneer.

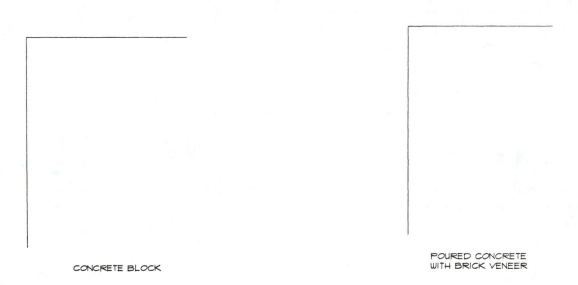

CONCRETE BLOCK

POURED CONCRETE
WITH BRICK VENEER

PROBLEM 11-4

Using the drawing below as a guide, draw and label the required information needed to show a two-story foundation made of concrete block. Thicken as required to support brick veneer.

CONCRETE BLOCK
WITH BRICK VENEER

PROBLEM 11-5

Place the needed dimension and extension lines. Place XX'-XX" to represent where dimension lines should be placed.

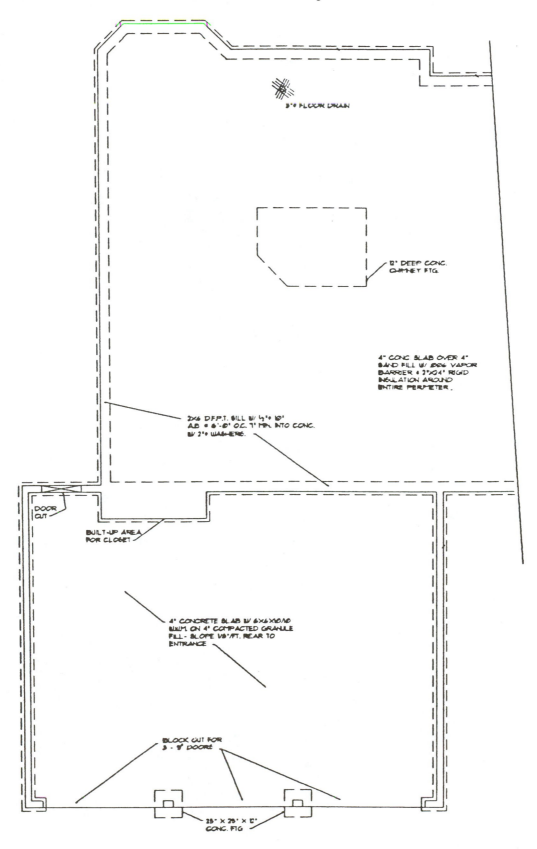

PROBLEM 11-6

Place the needed dimension and extension lines. Place XX'-XX" to represent where dimension lines should be placed.

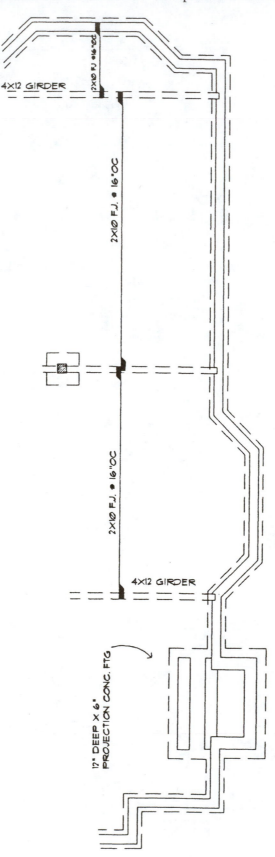

PROBLEM 11-7

Using the drawing below, draw, label, and completely dimension a plan view for a masonry footing and a 5'-0" x 2'-8" chimney. Assume the chimney extends 24" out from the exterior face of a one-story stem wall.

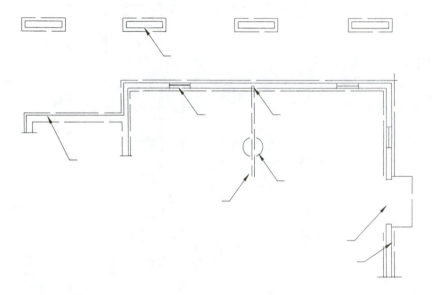

PROBLEM 11-8

Dimension and label the foundation plan below. Provide complete notes for each component. Assume: Standard sizes for all materials except for a 4 x 10 girder, 18" dia. x 8" piers, 30" x 18" crawl access, 24" x 6" vents.

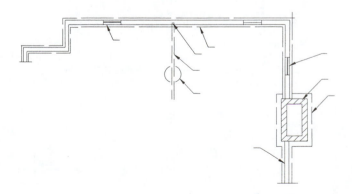

PROBLEM 11-9

Identify five dimensioning errors in the drawing below.

a. _____

b. _____

c. _____

d. _____

e. _____

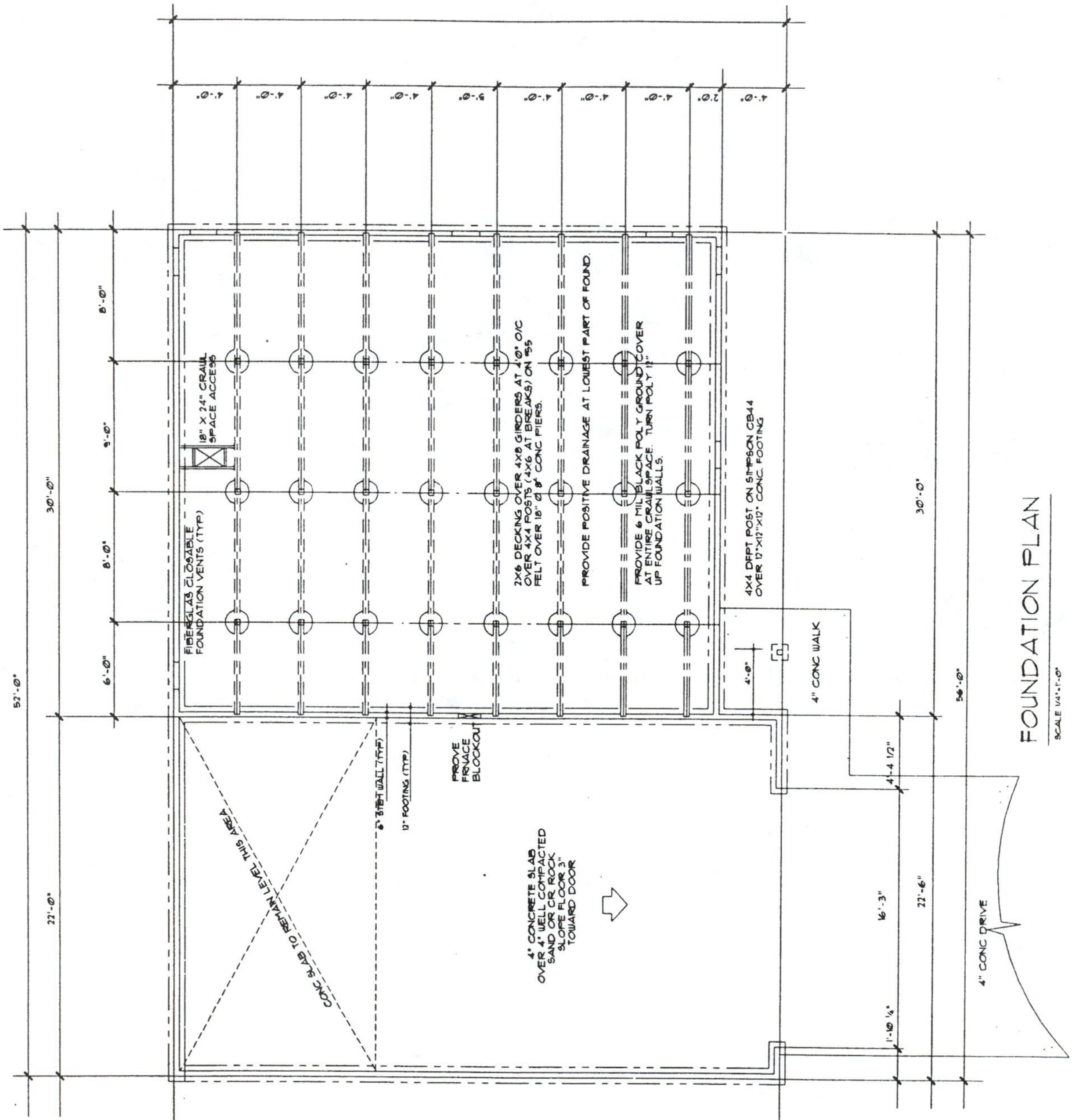

FOUNDATION PLAN
SCALE 1/4"=1'-0"

PROBLEM 11-10

Identify five errors in construction or drawing method in the plan below.

a. _____

b. _____

c. _____

d. _____

e. _____

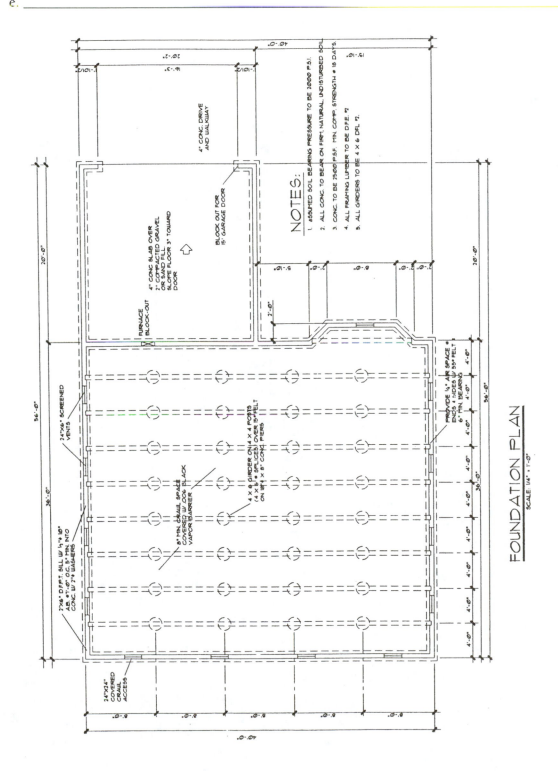

FOUNDATION PLAN

SCALE: 1/4" = 1'-0"

PROBLEM 11-11

Label and dimension the foundation plan below with the following information. Stem walls to be concrete block. Show a 1' x 6' deep cantilever for a bay window. Use 2 x 10 floor joists at 24" o.c. 6 x 8 girder w/supports @ 4'-6" o.c. Provide a sunken floor as indicated. Show a 30" x 18" crawl access, vents, and vapor barrier.

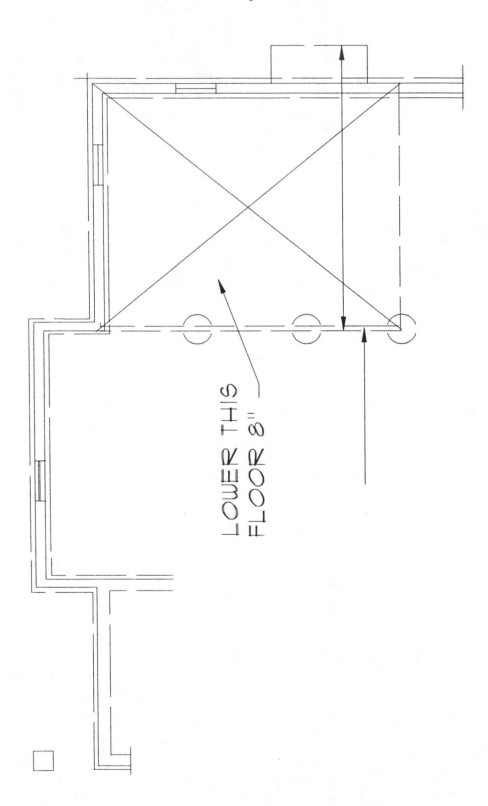

LOWER THIS FLOOR 8"

PROBLEM 11-12

On a separate sheet of paper, draw a 10' x 15'-wide portion of a concrete slab. Locate a 3-inch diameter floor drain 6' in and 8' down from the left edge. Show an area 8' x 9' sunken 7 1/2". Specify a 10" x 3" g.i. duct somewhere in the step. Provide dimensions and notes as required by the material.

PROBLEM 11-13

On a separate sheet of paper, draw a portion of a foundation plan with a post-and-beam foundation floor system. Show a step in the floor 10' from a stem wall made of concrete blocks. Provide notes and dimensions as required by the material.

PROBLEM 11-14

On a separate sheet of paper, draw a portion of a foundation plan using 2 x 6 floor joists @ 24" o.c. spanning between a concrete block stem wall and a 4 x 8 girder with 18" x 8"-deep concrete piers. Place the girder 9'-0" from the stem wall. Provide an access, required vents, and other materials normally associated with a joist floor system. Provide notes and dimensions as required by the material.

PROBLEM 11-15

Draw a plan view showing a portion of a concrete slab intersecting a wood floor framed with 2 x 8 floor joists @ 16" o.c. Support the floor joists on a 4 x 8 girder on 18" diameter concrete piers 10' away from the concrete slab. Have the floor joists extend 24" past the girder. Place complete notes and dimension lines to locate all needed materials.

PROBLEM 11-16

Identify ten construction errors in the following foundation specifications.

a. _____   f. _____
b. _____   g. _____
c. _____   h. _____
d. _____   i. _____
e. _____   j. _____

FOUNDATION NOTES

1. FOOTINGS ARE TO BEAR ON UNDISTURBED LEVEL SOIL DEVOID OF ANY ORGANIC MATERIAL AND STEPPED AS REQUIRED TO MAINTAIN THE REQUIRED DEPTH BELOW THE FINAL GRADE.

2. ASSUMED SOIL BEARING PRESSURE OF 2,000 PSI

3. ANY FILL UNDER GRADE SUPPORTED SLABS TO BE A MINIMUM OF 2" GRANULAR MATERIAL COMPACTED TO 95%

4. CONCRETE TO DEVELOPE A MIN. OF 2500 PSF @ 18 DAYS WITH A MIN. OF 5 SACKS OF CEMENT PER YARD AND A MAXIMUM SLUMP OF 4".

5. CONCRETE SLABS TO HAVE CONTROL JOINTS AT 25' (MAXIMUM) INTERVALS EA. WAY.

6. CONCRETE SIDEWALKS TO HAVE ¾" TOOLED JOINTS AT 5' O.C. (MINIMUM) SLABS TO BE 3" MIN. THICK.

7. REINFORCING STEEL TO BE A-615 GRADE 40. WELDED WIRE MESH TO BE A-185.

8. EXCAVATE THE SITE TO PROVIDE A MIN. OF 15" CLEARANCE UNDER ALL GIRDERS.

9. COVER CRAWL SPACE W/ 4 MIL BLACK VISQUEEN AND EXTEND UP FDTN. WALLS TO P.T. MUDSILL.

10. PROVIDE A MINIMUM OF .5 SQ. FT. OF VENTILATION AREA FOR EACH 150 SQ. FT. OF CRAWLSPACE AREA. VENTS ARE TO BE CLOSABLE WITH ¼" OPENINGS IN CORROSIVE RESISTANT SCREEN. POST NOTICE RE: OPENING VENTS AT THE ELECTRICAL PANEL.

11. ALL WOOD IN CONTACT WITH CONCRETE TO BE PRESSURE TREATED OR PROTECTED WITH 55# ROLL ROOFING.

12. BEAM POCKETS IN CONCRETE TO HAVE 3/4" AIRSPACE @ SIDES AND ENDS W/ A MINIMUM BEARING OF 2".

13. PROVIDE CRAWLSPACE DRAIN AS PER SEC. 2910 OF UBC.

14. WATERPROOF BASEMENT WALLS BEFORE BACKFILLING PROVIDE A 2" Ø PERFORATED DRAIN TILE BELOW THE TOP OF THE FOOTING (SEE BUILDING SECTIONS).

DIRECTIONS

1. Refer to the floor plans in Chapter 4 to determine dimensions and load-bearing wall locations.
2. Use the same scale to draw the foundation plan that was used to draw the floor plan.
3. Use the drawing of the foundation plan for reference only. Refer to reading materials and lecture notes for complete information. Provide complete dimensions and specifications to complete the drawing.
4. When the drawing is complete, turn a copy in to your instructor for evaluation.

PROBLEM 11-17(A)

Lay out a foundation plan using a post-and-beam floor system. Use 4 x 6 girders on 18" diameter x 8" piers unless noted.

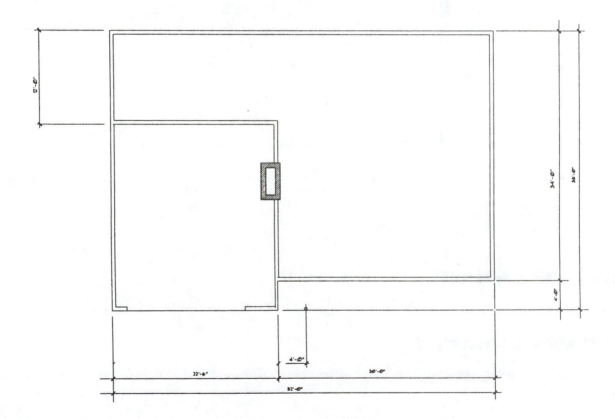

PROBLEM 11-17(B)

Lay out a foundation plan using a joist floor system. Use 2 x 8 floor joists @ 16" o.c. supported on 6 x 10 girders on 21" diameter x 12"-deep piers.

PROBLEM 11-17(C)

Lay out a foundation plan using a concrete slab floor system. Specify wire mesh and 2" rigid insulation under the entire slab.

PROBLEM 11-18(A)

Lay out a foundation plan using a joist floor system. Use 2 x 8 floor joists @ 16" o.c. supported on 4 x 10 girders. Space girders @ 14' o.c. and support on 21" diameter x 12"-deep piers @ 6'-0" o.c.

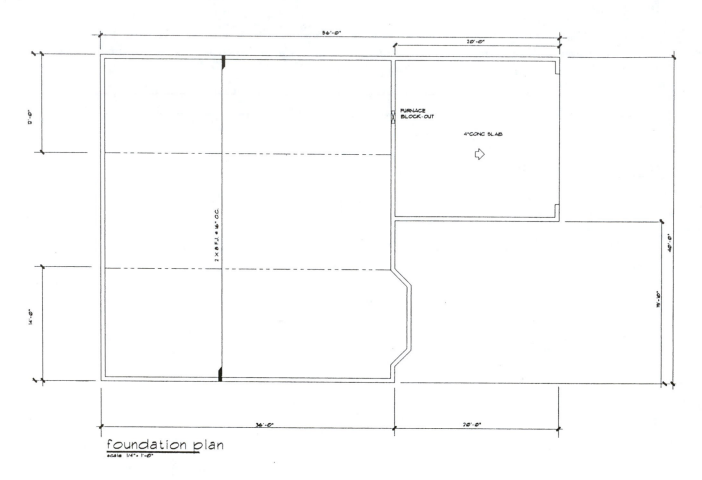

foundation plan
scale 1/4" = 1'-0"

PROBLEM 11-18(B)

Lay out a foundation plan using a post-and-beam floor system. Use 4 x 6 girders on 18" diameter x 8" piers.

PROBLEM 11-18(C)

Lay out a foundation plan using a concrete slab floor system. Specify wire mesh and 2" rigid insulation under the entire slab.

PROBLEM 11-19(A)

Lay out a foundation plan using a post-and-beam floor system. Use 4 x 8 girders on 18" diameter x 8" piers.

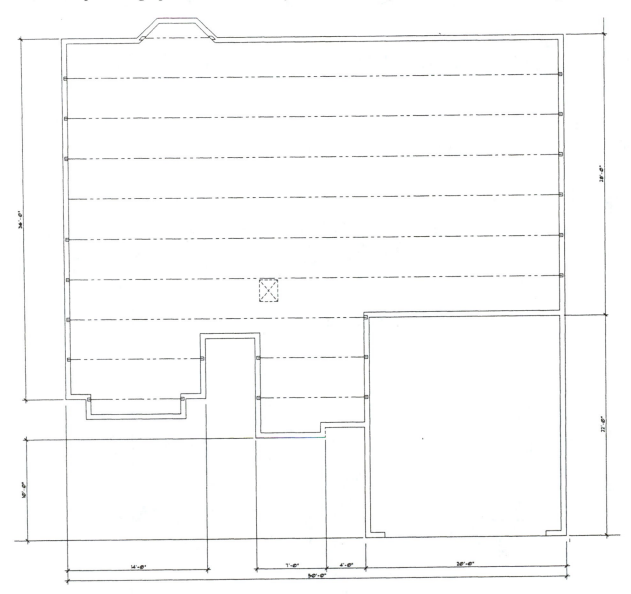

PROBLEM 11-19(B)

Lay out a foundation plan using a joist floor system. Use 2 x 10 floor joists @ 16" o.c. on 4 x 12 girders on 20" x 20" x 8" concrete piers.

PROBLEM 11-20

Lay out a foundation plan using a joist floor system. Use 6 x 10 girders under load-bearing walls (girders 2, 3, 4) and 4 x 10 girders to support floor loads (girders 1, 5). Use 22" diameter x 10" concrete piers @ 6'-0" o.c. unless noted.

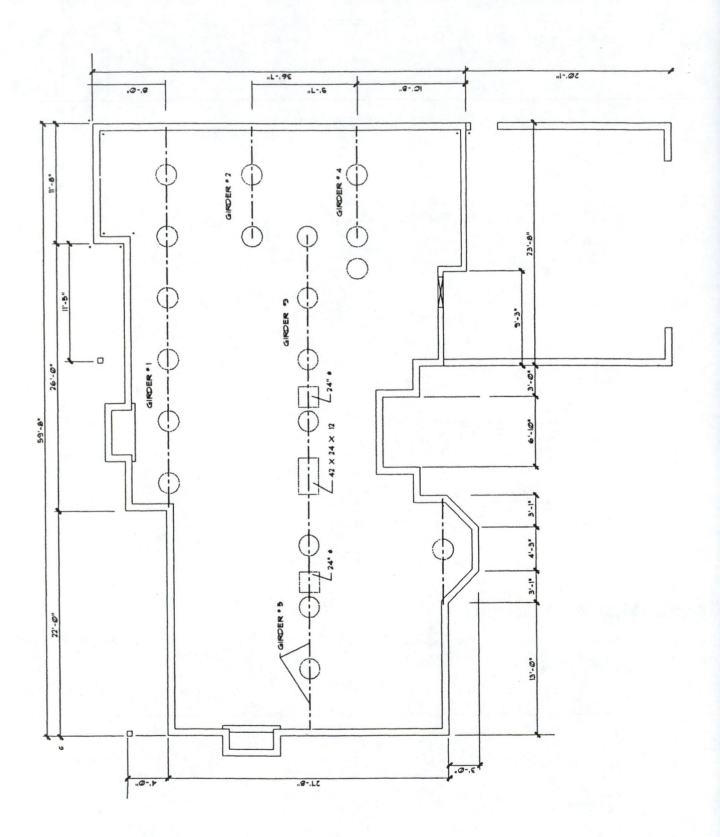

PROBLEM 11-21

Lay out a foundation plan using a post-and-beam floor system. Assume standard sizes for girders and piers unless noted.

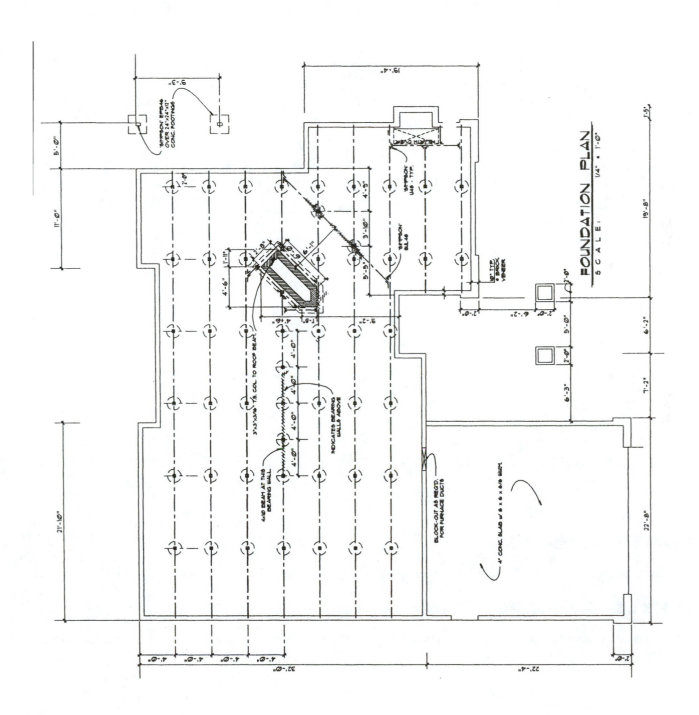

PROBLEM 11-22(A)

Lay out a foundation plan using a concrete slab floor system. Specify wire mesh and 2" rigid insulation under the entire slab.

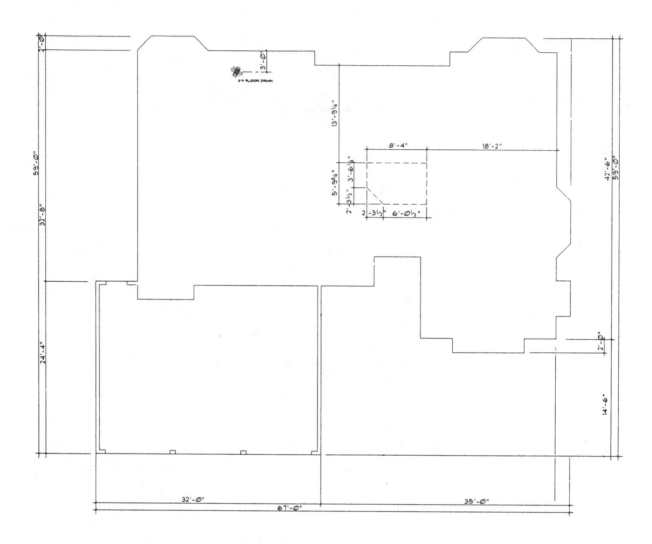

PROBLEM 11-22(B)

Lay out a foundation plan using a post-and-beam floor system. Assume standard sizes for girders and piers.

PROBLEM 11-22(C)

Lay out a foundation plan using a joist floor system. Use 2 x 10 floor joists @ 16" o.c. supported by 4 x 12 girders on 30" square concrete piers.

PROBLEM 11-23

Lay out a foundation plan using a joist floor system. Use 4 x 12 girders and 2 x 10 floor joists.

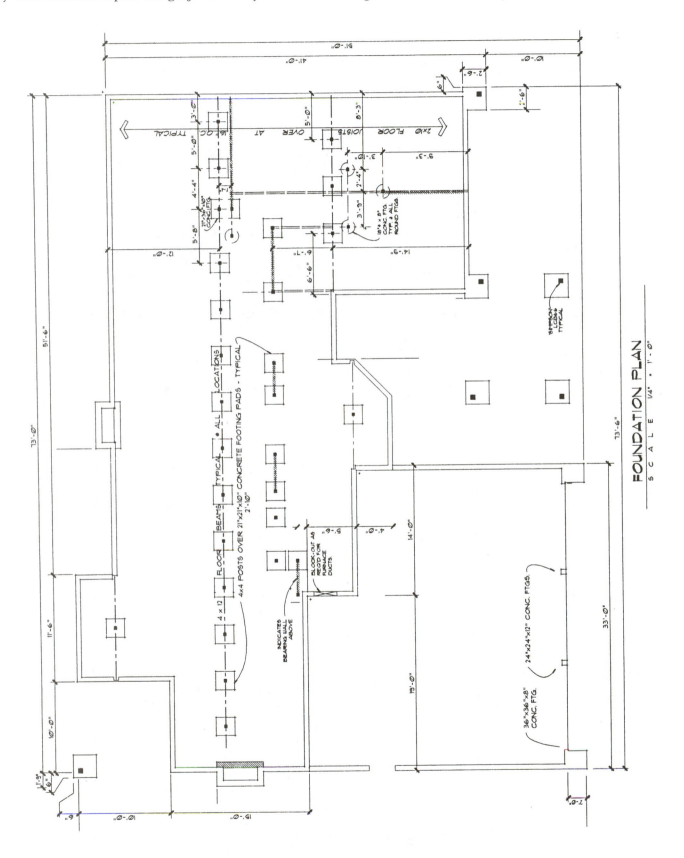

FOUNDATION PLAN
SCALE 1/4" = 1'-0"

PROBLEM 11-24

Lay out a foundation plan using a post-and-beam floor system. Determine the size and location of girders and piers based on reading material. Provide an alternate foundation plan for a concrete slab

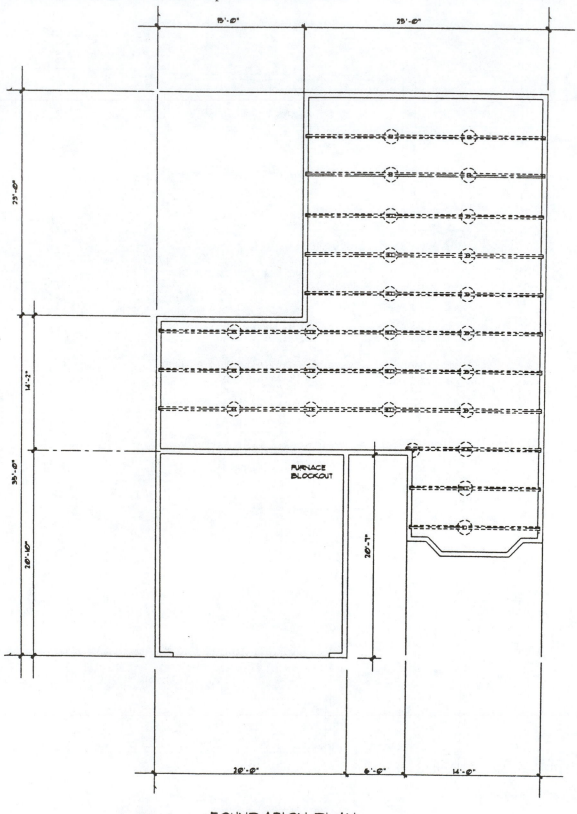

FOUNDATION PLAN
SCALE 1/4" = 1'-0"

PROBLEM 11-25

Lay out a foundation plan using a joist floor system. Determine the floor joist size using span tables for lumber suitable for your area.

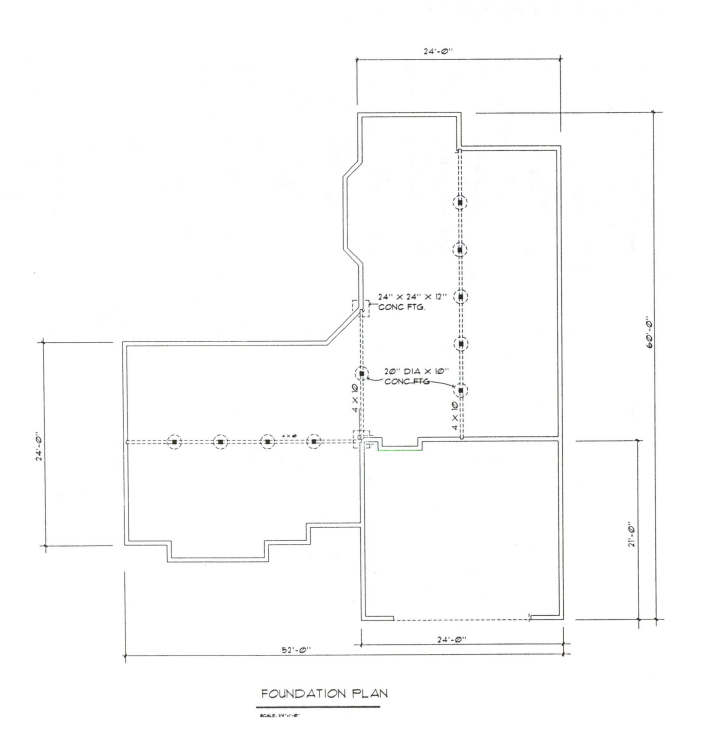

FOUNDATION PLAN

SCALE: 1/4"=1'-0"

PROBLEM 11-26

Use the floor plan from Chapter 4, the attached preliminary design, and the guidelines in your text to layout and draw a foundation plan with a post and beam floor system.

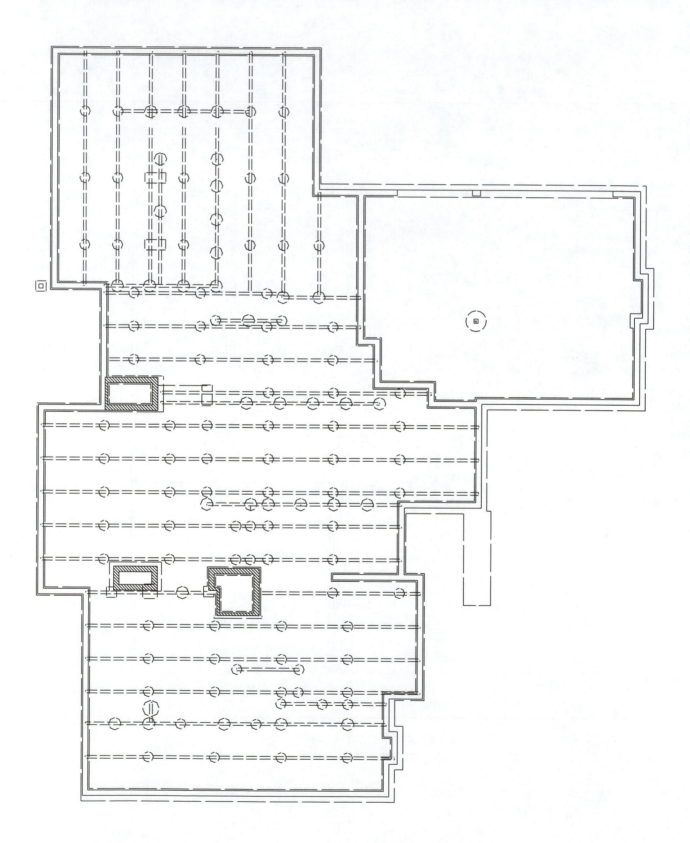

PROBLEM 11-27

Use the floor plan from Chapter 4, the attached preliminary design, and the guidelines in your text to layout and draw a foundation plan with a post and beam floor system.

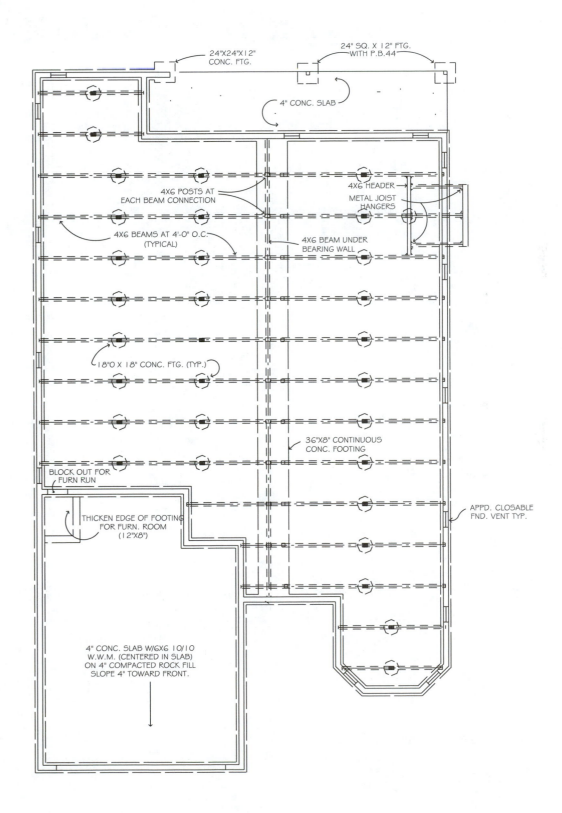

PROBLEM 11-28

Use the floor plan from Chapter 4, the attached preliminary design, and the guidelines in your text to layout and draw a foundation plan with a joist floor system.

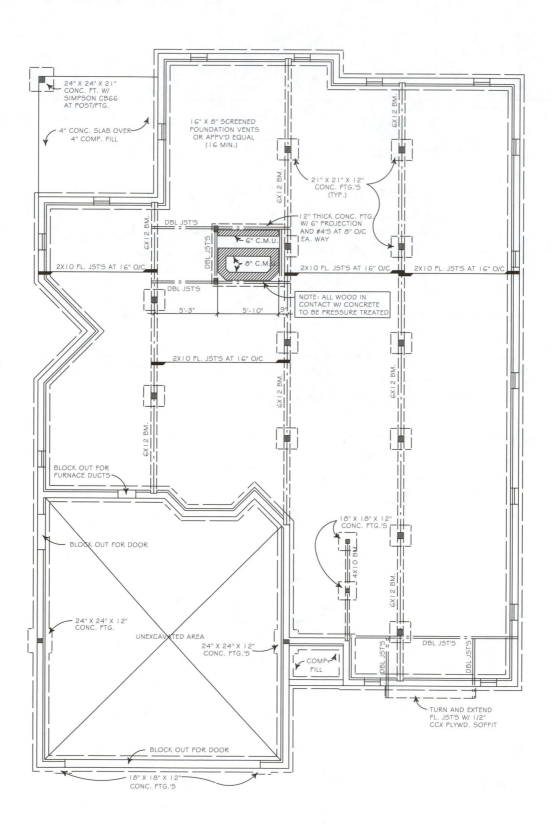

24" X 24" X 21"
CONC. FT. W/
SIMPSON CB66
AT POST/FTG.

4" CONC. SLAB OVER
4" COMP. FILL

16" X 8" SCREENED
FOUNDATION VENTS
OR APPV'D EQUAL
(16 MIN.)

6X12 BM.

6X12 BM.

21" X 21" X 12"
CONC. FTG.'S
(TYP.)

DBL JST'S

12" THICK CONC. FTG.
W/ 6" PROJECTION
AND #4'S AT 8" O/C
EA. WAY

6X12 BM.

DBL JST'S

6" C.M.U.

8" C.M.

2X10 FL. JST'S AT 16" O/C

2X10 FL. JST'S AT 16" O/C

2X10 FL. JST'S AT 16" O/C

DBL JST'S

NOTE: ALL WOOD IN
CONTACT W/ CONCRETE
TO BE PRESSURE TREATED

5'-3"

5'-10"

9"

2X10 FL. JST'S AT 16" O/C

6X12 BM.

6X12 BM.

BLOCK OUT FOR
FURNACE DUCTS

18" X 18" X 12"
CONC. FTG.'S

BLOCK OUT FOR DOOR

4X10 BM.

6X12 BM.

24" X 24" X 12"
CONC. FTG.

UNEXCAVATED AREA

24" X 24" X 12"
CONC. FTG.'S

COMP.
FILL

DBL JST'S

DBL JST'S

DBL JST'S

TURN AND EXTEND
FL. JST'S W/ 1/2"
CCX PLYWD. SOFFIT

BLOCK OUT FOR DOOR

18" X 18" X 12"
CONC. FTG.'S

PROBLEM 11-29

Use the floor plan from Chapter 4, the attached preliminary design, and the guidelines in your text to layout and draw a foundation plan with a joist floor system.

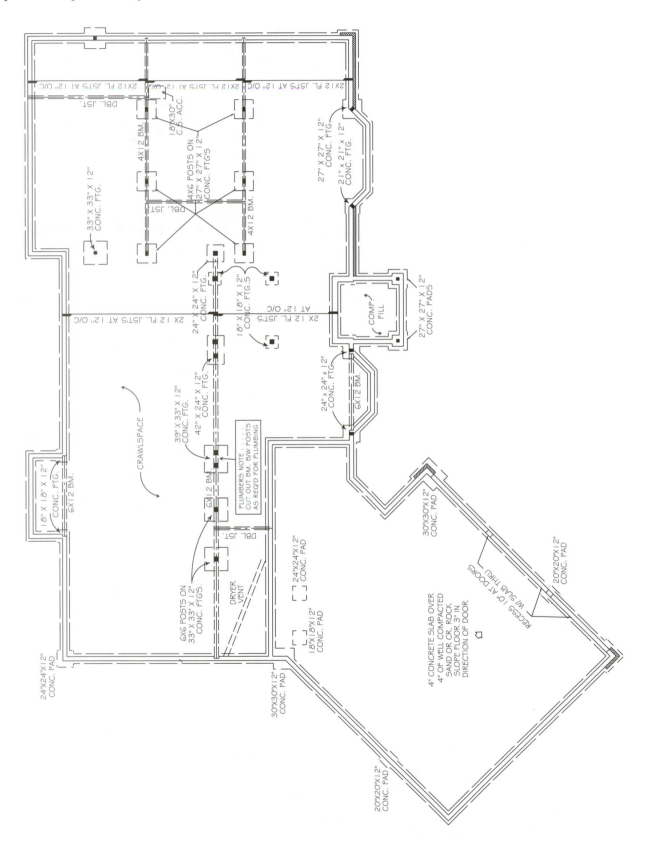

PROBLEM 11-30

Use the floor plan from Chapter 4, the attached preliminary design, and the guidelines in your text to layout and draw a foundation plan with a joist floor system.

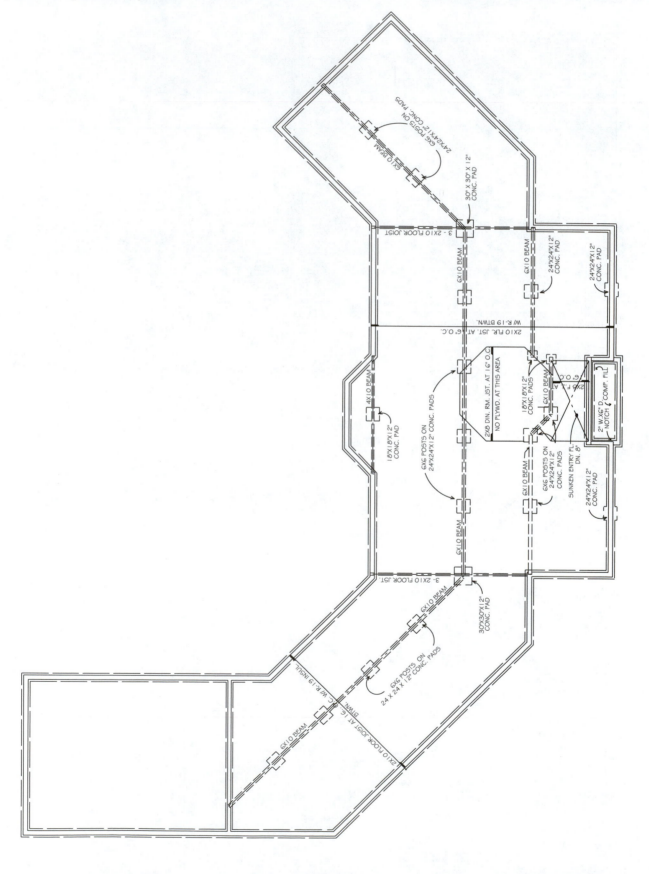

PROBLEM 11-31

Lay out a foundation plan for a two-story residence using a post-and-beam floor system. Determine the size and location of girders and piers based on reading material.

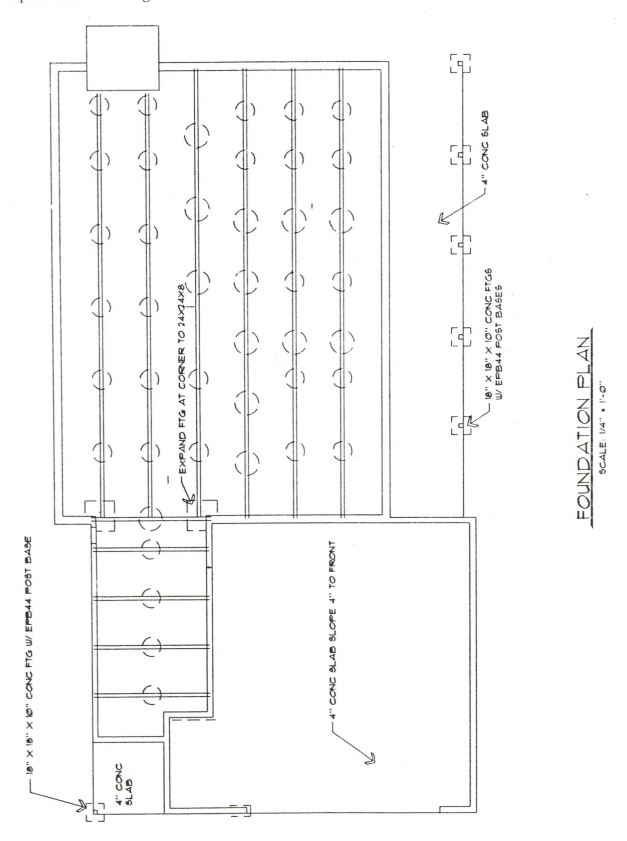

FOUNDATION PLAN

SCALE: 1/4" = 1'-0"

4" CONC SLAB

18" X 18" X 10" CONC FTGS
W/ EPB44 POST BASES

EXPAND FTG AT CORNER TO 24X24X8

18" X 18" X 10" CONC FTG W/ EPB44 POST BASE

4" CONC SLAB SLOPE 4" TO FRONT

4" CONC SLAB

CHAPTER
Details

DIRECTIONS: Unless other instructions are given by your instructor, complete the following details for a one-level house. Use a scale of 3/4" = 1'-0" with proper line weight and quality. Provide notes required to label typical materials. Provide dimensions based on common practice in your area. Omit insulation unless specified.

PROBLEM 12-1

Show a typical foundation with poured concrete with a 2 x 4 key, 2 x 6 f.j. at 16" o.c. x 4 studs at 24" o.c.

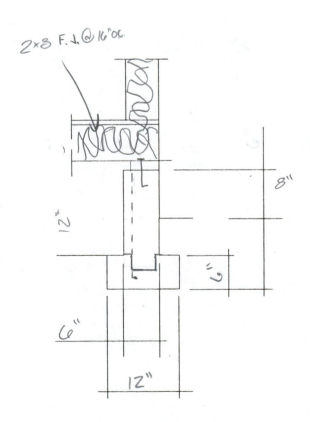

PROBLEM 12-2

Show a typical post-and-beam foundation system. Show 4 x 6 girders @ 48" o.c. parallel to the stem wall.

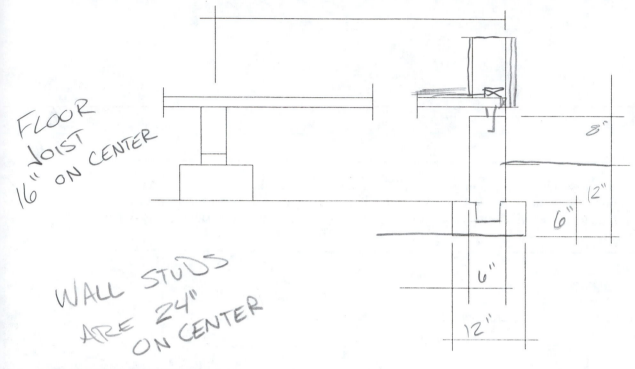

FLOOR
JOIST
16" ON CENTER

WALL STUDS
ARE 24"
ON CENTER

8"

12"

6"

6"

12"

PROBLEM 12-3

Complete the drawing showing a stem wall using concrete blocks with #4 steel @ 18" o.c. each way. Use 2 x 8 f.j. at 24" o.c. and 2 x 4 studs at 24" o.c.

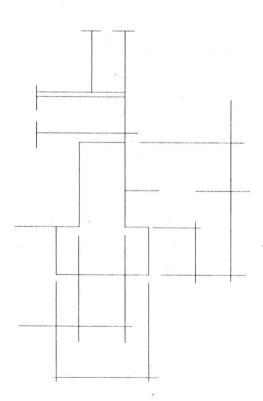

PROBLEM 12-4

Draw a foundation showing a typical foundation for mono-poured concrete with 2 x 8 f.j. at 16" o.c. and 2 x 4 studs at 24" o.c. Thicken footing as required for brick veneer over 1/2" waferboard and Tyvek.

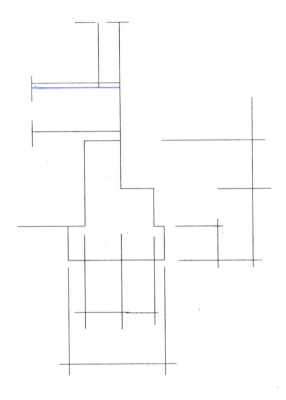

PROBLEM 12-5

Draw a foundation showing a concrete floor system. Use #10 x 10-4" x 4" WWM 2" dn. from top surface and 2-#4 bars centered in the footing 2" up and down. Use 2 x 4 studs at 16" o.c. for walls.

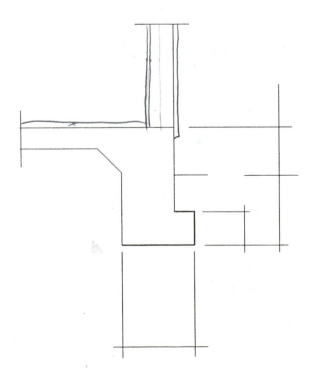

PROBLEM 12-6

Draw a detail showing an interior footing for a concrete slab supporting a 2 x 4 stud-bearing wall. Anchor the wall with Ramset type fasteners (or equal). Use the same slab reinforcing that was used in Problem 12-5.

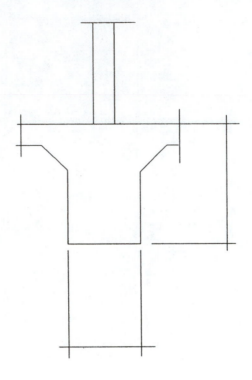

DIRECTIONS: Complete problems 12–7 through 12–16 using a scale of 1/2" = 1'-0".

PROBLEM 12-7

Complete the detail to show a change in floor elevation. Floor step to be 7 3/4" maximum, with 4 x 10 girder at the step, 2 x 10 f.j. at standard spacing, 3/4" ply, Metal hangers by Simpson. Use 18" dia. x 8" pier.

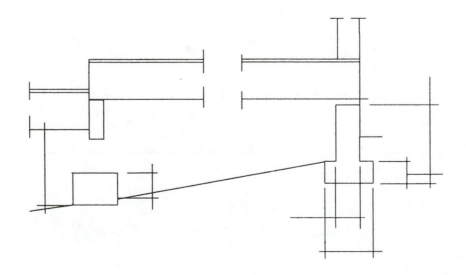

PROBLEM 12-8

Draw a detail to show the intersection of a concrete floor supported on a 8" x 8" x 16" concrete block stem wall with floor joists. Show the slab with #10 x 10-4" x 4" WWM 2" dn. from top surface. Reinforce the block wall with #5 @ 48" o.c. Use 2 x 10 f.j. at 16" o.c. Attach to wall with 2 x 10 D.F.P.T. ledger with 5/8" a.b. @ 32" o.c. staggered 3" up and dn. ledger. Use 5/8" plywood sheathing and set flush to slab. Support the floor with a 6 x 10 girder on a 4 x 6 post, 21" dia. piers. Cantilever the floor joist 12" past the girder.

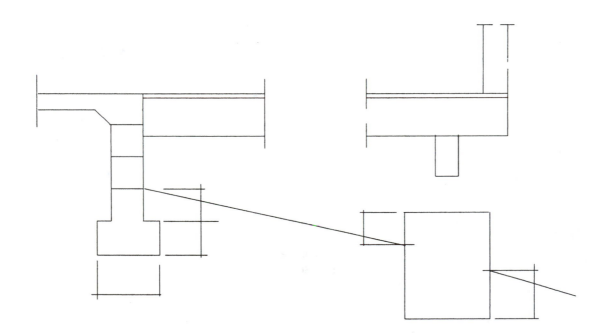

PROBLEM 12-9

Design a retaining wall for an 8'-tall basement using poured concrete. Use 2 x 10 f.j. perpendicular to the wall, w/5/8" dia. a.b. @ 32" o.c. Show #4 @ 18" o.c. each way @ 2" from tension side with 2-#5 in footing 2" up. Use a scale of 1/2" = 1'-0".

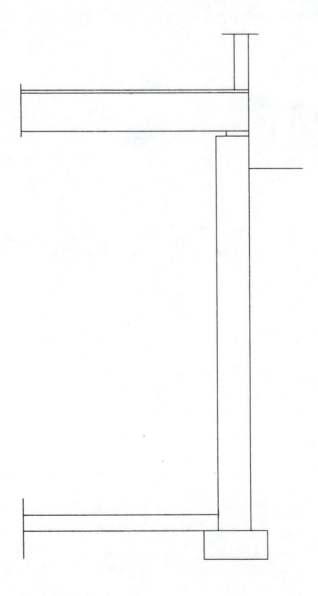

PROBLEM 12-10

Show the top portion of a retaining wall using poured concrete. Use 2 x 10 f.j. perpendicular to the wall, w/5/8" dia. a.b. @ 24" o.c. Show support for brick veneer over 1/2" waferboard.

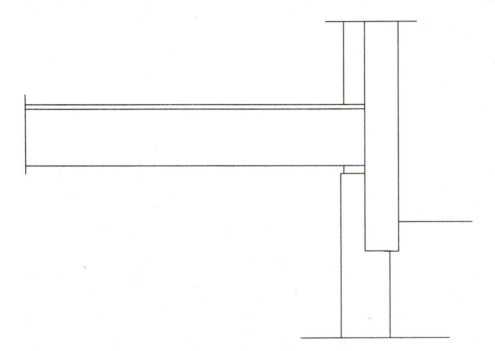

PROBLEM 12-11

Show 2 x 10 f.j. @ 12" o.c. cantilevered 12" past a 5 1/8 x 13 1/2 Fb 2200 glu-lam beam. Anchor joists to beam with Simpson Co. joist anchors. Support the beam on a 6 x 6 post using an appropriate Simpson Co. column cap. Show a 3/8" dia. eyebolt 8" below the top of the post. Use 3/16" steel cable each way, with 2" between each eyebolt. Show an end and side view.

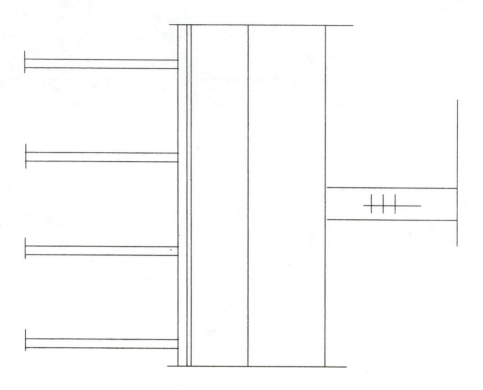

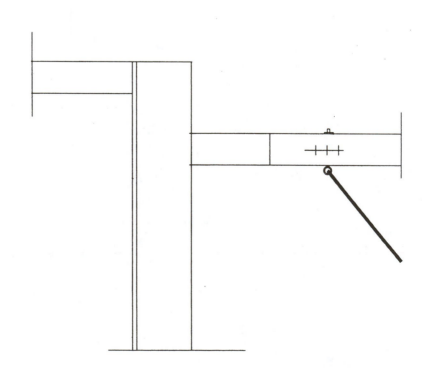

PROBLEM 12-12

Use a scale of 1/2" = 1'-0" to show a 6 x 6 post intersecting a 21" dia. x 36" deep concrete pier. Use an appropriate column base. Show a 3/8" dia. eyebolt with 1 1/2" dia. washers each side 8" above the bottom of the post. Use 3/16" steel cable each way, with 2" between each eyebolt. Use 3-#4 each way 2" up from the bottom of footing. Show end and side views.

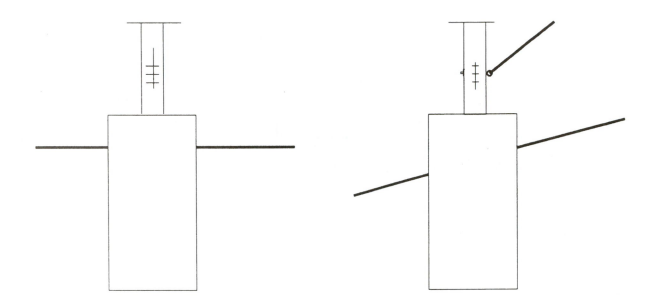

PROBLEM 12-13

Use a scale of 1/2" = 1'-0" to show an energy-efficient wall/foundation intersection with a concrete slab. Show all insulation and caulking. Show a 2 x 6 P.T. sill let-into slab with 2" rigid insulation on the exterior side of the concrete.

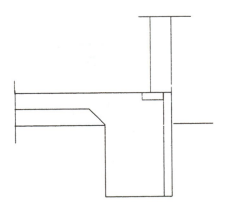

PROBLEM 12-14

Using a scale of 1/2" = 1'-0", show an energy-efficient wall/foundation intersection. Use a concrete slab with a thermal break between the slab and the footing. Show all insulation and caulking. Show slab with 2" rigid insulation on the exterior side and below the concrete.

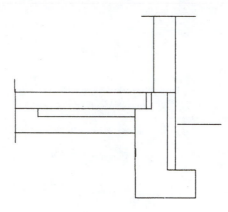

PROBLEM 12-15

Using a scale of 1/2" = 1'-0", show an energy-efficient wall/foundation intersection. Use a post-and-beam floor system and show all insulation and caulking. Provide 2" rigid insulation to keep the footing warm.

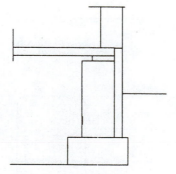

PROBLEM 12-16

Using a scale of 1/2" = 1'-0", show an energy-efficient wall/foundation intersection. Use a joist floor system with 2 x 6 f.j. @ 16" o.c. Show all caulking and insulation on the exterior side of the stem wall.

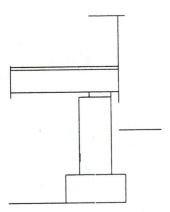

PROBLEM 12-17

Using a scale of 1/2" = 1'-0", show a floor with a heated plunum below. Frame the floor with 2 x 6 f.j. @ 16" o.c. (no insulation). Use a 24" crawl space covered with a 3 1/2" slab over 4" sand fill and 2" rigid insulation (specify a type). Show all caulking and insulation on the interior side of the stem wall.

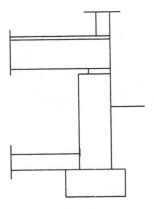

PROBLEM 12-18

Design a detail to show a 30" change in floor elevation. Use a 4 x 10 girder at the step, 2 x 10 f.j. at standard spacing, 3/4" ply. Use 18" dia. x 12" pier. Show a rail formed with dec. oak spindles and a 2 x 6 handrail with decorative molding. Use a scale of 1/2" = 1'-0".

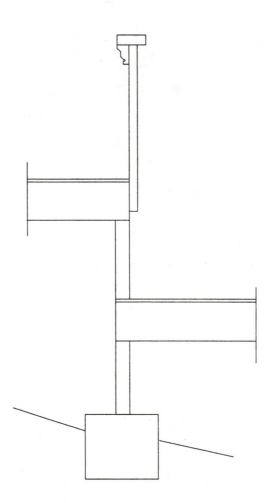

PROBLEM 12-19

Draw a detail to show the intersection of a concrete floor supported on a poured concrete stem wall with floor joists. Show the slab w/#10 x 10-4" x 4" WWM 2" dn. from top surface. Reinforce the stem wall w/#5 @ 32" o.c. and extend 18" into slab. Show #5-3" down and 3" up in slab and footing. Use 2 x 8 f.j. at 12" o.c. Attach to wall with 2 x 8 D.F.P.T. ledger w/5/8" a.b. @ 32" o.c. staggered 3" up and dn. ledger. Specify a Simpson joist hanger (be specific by model number). Use 3/4" ply sheathing and set flush to slab. Cantilever the floor joist 18" past the girder.

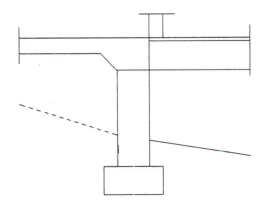

PROBLEM 12-20

Show a detail of the intersection of a 48"-high concrete restraining wall. Use a 30" -wide footing with 2-#5 cont. 3" up @ 12" o.c. and #5 @ 24" o.c. in wall 2" from tension side. Use 4" drain in 8" x 30" gravel bed. Use 2 x 6 sill with 5/8" a.b. @ 24" o.c. Use an 8' ceiling with 2 x 10 f.j. @ 16" o.c. Cantilever 12" past wall w/1" exterior stucco. Use a scale of 1/2" = 1'-0".

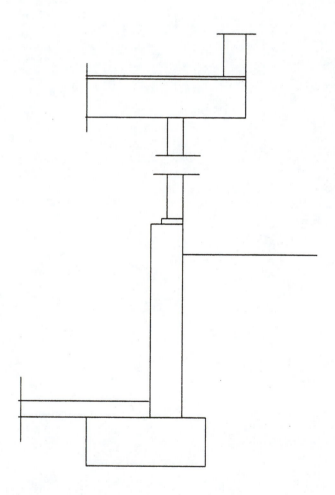

PROBLEM 12-21

Show a detail of the intersection of a 60"-high concrete restraining wall. Use a 30"-wide footing with 3-#5 cont. @ 12" o.c. 3" up and #5 @ 18" o.c. in wall 2" from tension side both ways. Use a 6 x 6 key 6" from footing toe. Use 4" drain in 8" gravel bed full height. Use 2 x 6 sill with 5/8" a.b. @ 24" o.c. Use a scale of 1/2" = 1'-0".

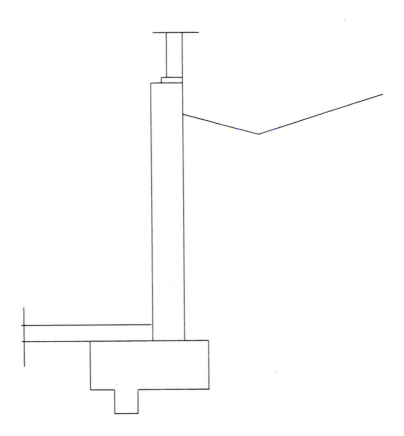

PROBLEM 12-22

Show the intersection of a 2 x 10 joist floor with a 2 x 6 exterior wall and a wood deck. Support the deck with 2 x 8 floor joists, 2 x 4 decking laid flat w/1/4" spacing. Place the deck 1" below the floor. Show a ledger with joist hangers and a 4 x 10 beam. Use a 2 x 2 vertical railing. Use a scale of 1/2" = 1'-0".

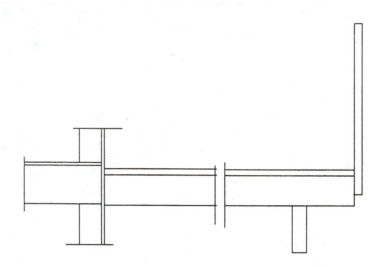

PROBLEM 12-23

Show the intersection of a concrete garage slab with a concrete block stem wall supporting 2 x 6 floor joists. Flash between the wood and concrete with 26 ga. flashing. Use a scale of 1/2" = 1'-0".

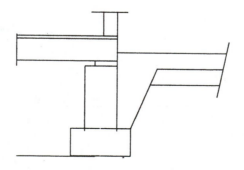

PROBLEM 12-24

Show the intersection of a 3"-dia. steel column with 12"-dia. 1-level footing below a 4" concrete slab. Support the column on a 1/2" x 8" x 8" steel plate. Set the plate on 1" epoxy cement. Use (4) 1/2" Ø a.b. 1 1/2" from each edge of the plate. Weld the column to the plate with a 3/16 fillet weld, all around, field weld. Use WWM in the slab and (3)-#4 dia. each way 2" up from the bottom of the footing. Use a scale of 1/2" = 1'-0".

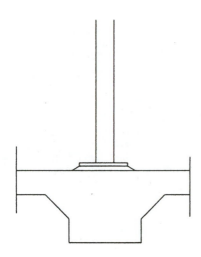

PROBLEM 12-25

Show a detail of a W10 x 49 steel wide flange supporting 2 x 10 floor joists. Use a 2 x 6 ledger bolted with 1/2" stud @ 32" o.c. staggered. Use solid blocking w/an A-35 connector at 16" o.c. staggered each side. Use a scale of 1 1/2" = 1'-0".

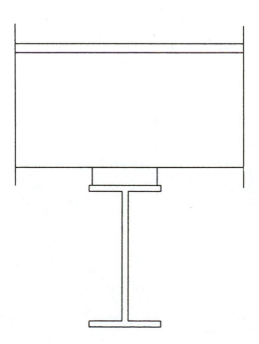

PROBLEM 12-26

Design a detail to show a 16" step in floor levels, 18" to the left of a 4 x 10 girder. Support the beam on 18" dia. piers. Use 2 x 8 f.j. @ 16" o.c.

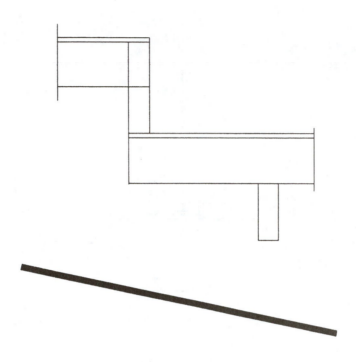

PROBLEM 12-27

Design a detail to show a 12" step in a post-and-beam floor system. Use 4 x 8 beams supported on 15" dia. piers. Use a scale of 1/2" = 1'-0".

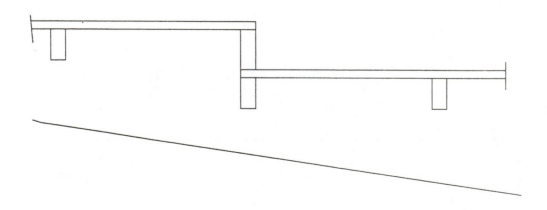

PROBLEM 12-28

Design a detail to show a 10" step in a concrete slab floor system at an interior 2-story load bearing wall. Use 5/8" a.b. @ 48" o.c. with #40 @ 3" from top and bottom. Use WWM in each slab. Use a scale of 1/2" = 1'-0".

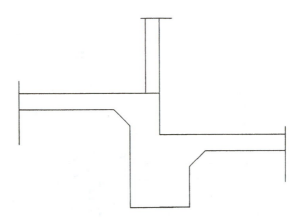

PROBLEM 12-29

Draw the intersection of a standard roof truss with a 2 x 6 stud wall. Truss to be standard with a 3/12 pitch, 300# comp roof with 24" overhang. Wall to be single wall construction, T1-11 plywood.

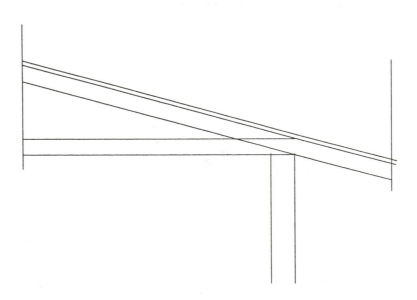

PROBLEM 12-30

Draw the intersection of a standard truss with a concrete block wall. The roofing is to be 5/12 pitch, covered with 235# comp. shingles. Box in the eave so that no tails are exposed. Vent with 2" wide continuous vent. The wall is to be reinforced with a #5 @ 32" o.c. each way. Use a bond beam at the top of the wall with (4)-#4 rebar. Solid grout all steel cells.

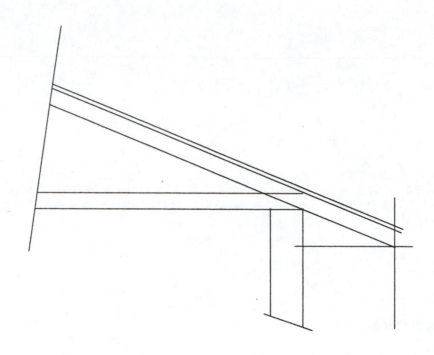

PROBLEM 12-31

Draw the intersection of a truss with a 2 x 6 wall. Trusses to be standard with a 6/12 pitch, cedar shakes. Walls to be double-wall construction with horizontal lap siding. Box in the eave so that no tails are exposed. Vent with 2"-wide continuous vent. Show a 2-2 x 12 header over a 2 x 6 nailer set at 6'-8" above the floor.

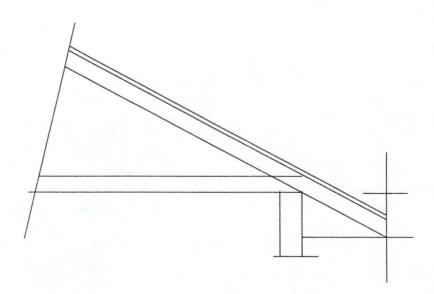

PROBLEM 12-32

Draw the intersection of a truss with a 2 x 4 wall. Truss to be scissor with a 5/12 exterior pitch, and 3/12 interior pitch with 235# comp roof. Use double-wall construction, with 1/2" x 8" bevel siding over 1/2" waferboard.

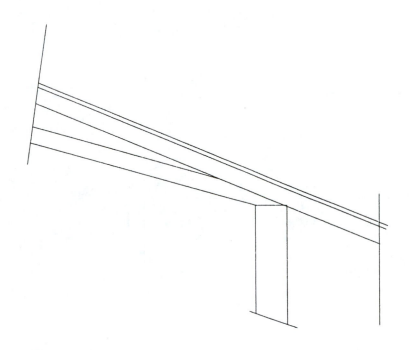

PROBLEM 12-33

Draw the intersection of a mono truss supported by a 6 x 14 beam. Truss to be @ 4/12 pitch. Show a 2 x 12 raft/c.j. intersecting the truss forming a ridge. Support 2 x 12 ledger with (3)-16d nails to each truss with U210 Simpson metal hanger. Provide required vents at the ridge and notch.

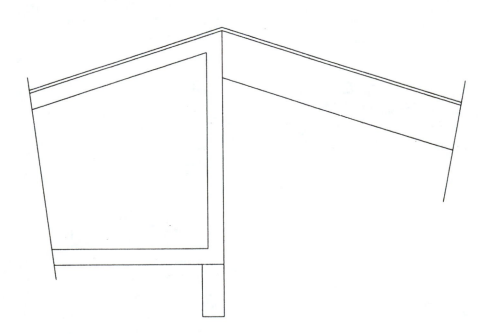

PROBLEM 12-34

Draw the intersection of a 2 x 6 rafter with a 2 x 4 wall. Roof pitch to be 5/12, 235# comp roofing, 2 x 6 ceiling joist. Wall to be single-wall construction, T1-11 ply. Box in the eave so that no tails are exposed. Vent with 2" wide continuous vent.

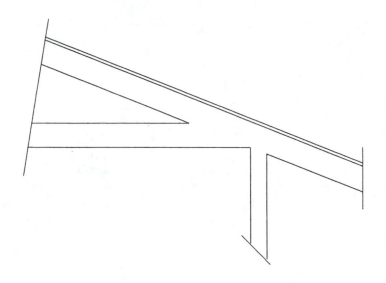

PROBLEM 12-35

Draw the intersection of a 2 x 8 rafter with a concrete-block wall. Roof to be 2/12 pitch with built-up roofing. Ceiling to be 2 x 6 at standard spacing. Wall to be reinforced with #5 @ 16" o.c.

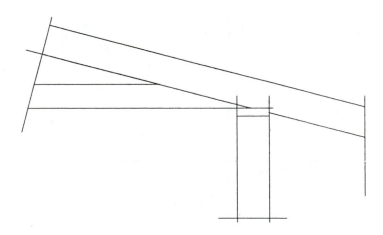

PROBLEM 12-36

Draw the intersection of a 2 x 12 rafter/ceiling joist with a 2 x 6 exterior wall. Roof to be 6/12 pitch, cedar shakes with 1/2" gyp bd. on interior side with R-30 batt insulation. Wall to be single-wall construction with 1" exterior stucco.

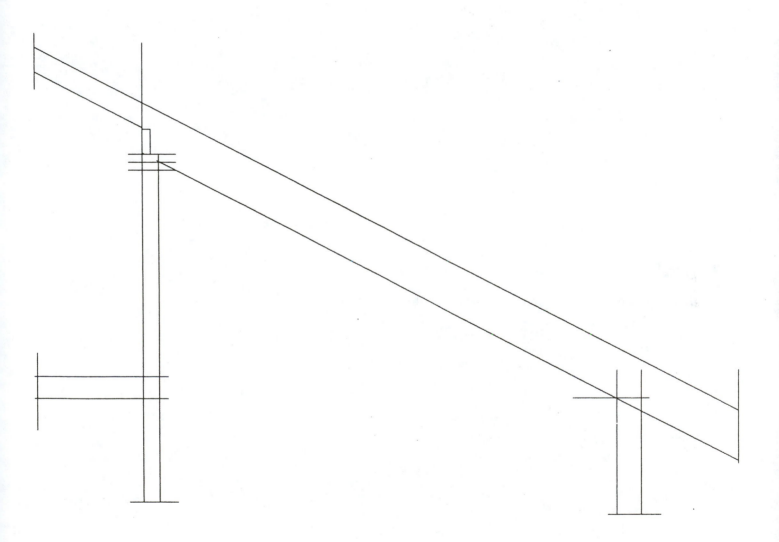

> **DIRECTIONS:** Use a scale of 3/4" = 1'-0" for drawings 12–37 through 12–39.

PROBLEM 12-37

Draw a detail showing the plan view of a 4 x 4 post in the center of a 16" x 16" stucco column. Frame the column with 2 x 4's.

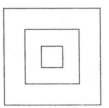

PROBLEM 12-38

Draw a detail showing the plan view of a 4 x 4 post in the center of a 16" x 16" brick column.

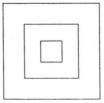

PROBLEM 12-39

Draw a detail showing a plan view of an exterior and interior corner using energy-efficient framing methods and R-21 insulation. Show all studs, caulking, and vapor barrier locations.

PROBLEM 12-40

Draw a detail using a scale of 1" = 1'-0" showing a wall intersecting a concrete slab. Use energy-efficient methods of framing and caulking. Use 2" x 24" "Dow Blue board" rigid insulation under the slab and on the back side of the stem wall.

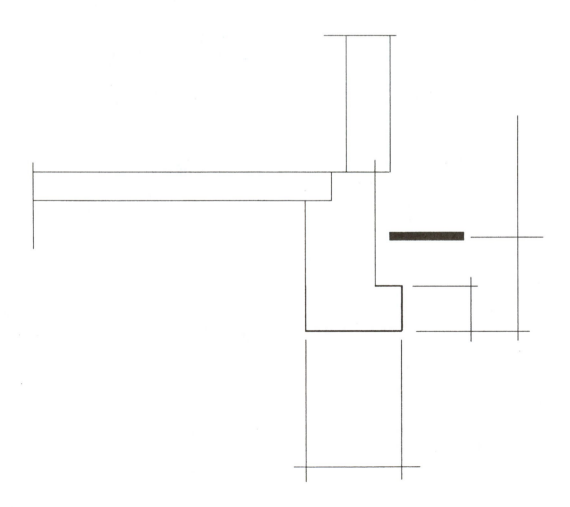

PROBLEM 12-41

Draw details using a scale of 3/4" = 1'-0" to contrast the difference between a standard truss and 2 x 12 rafter/ceiling joist intersecting a wall with double-wall construction. Insulate the roof to R-30 and show how eave vents will be protected from insulation. Indicate caulking locations.

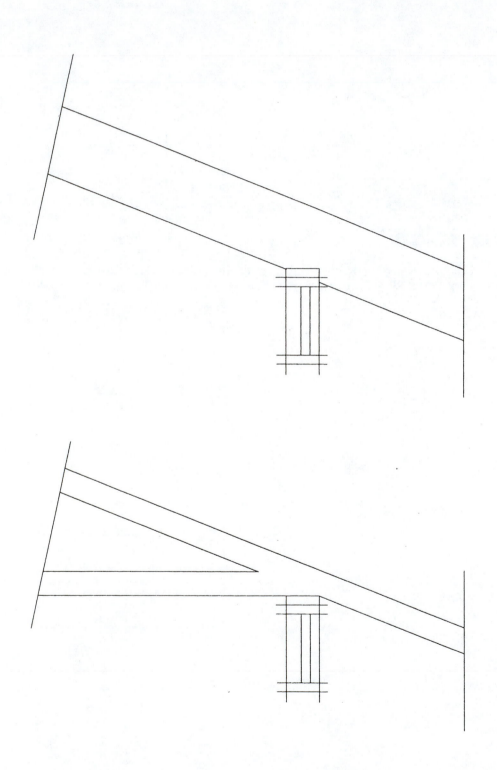

PROBLEM 12-42

Draw the intersection of a rafter with a 2 x 6 stud wall. Use a 3/12 pitch, 300# comp roof w/24" overhang. Wall to be single-wall construction, T1-11 ply (1/2" = 1'-0").

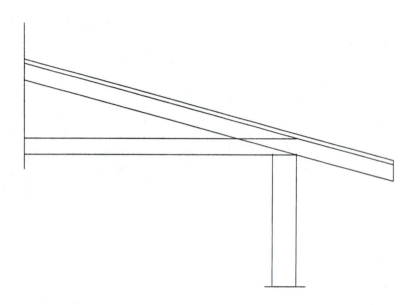

PROBLEM 12-43

Draw a detail showing a 5 1/8 x 15 glu-lam beam supported on a 6 x 6 column. Specify an appropriate column cap. Use 2 x 10 f.j. perpendicular to the glu-lam. Show a side and end view (1/2" = 1'-0").

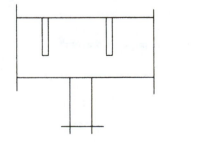 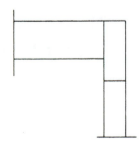

PROBLEM 12-44

Draw a detail showing a 5 1/8 x 15 glu-lam beam supported on a 3" dia. steel column. Specify an appropriate Simpson column cap. Weld cap to column w/3/16" fillet weld, all around, in field. Use 2 x 8 f.j. at 16" o.c. with metal joist hangers. Show side and end views (1" = 1'-0").

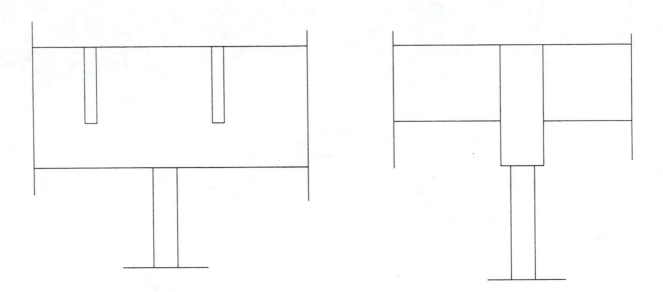

PROBLEM 12-45

Draw a detail of a concrete block parapet wall supporting a 2 x 10 ledger and 2 x 8 raft./c.j. @ 24" o.c. parallel to wall. Bolt ledger with 1/2" x 10" a.b. @ 32" o.c. staggered 2" up and down. Use 26 ga. flashing at the wall top and roof/wall intersection. Reinforce wall with #5 @ 32" o.c. each way.

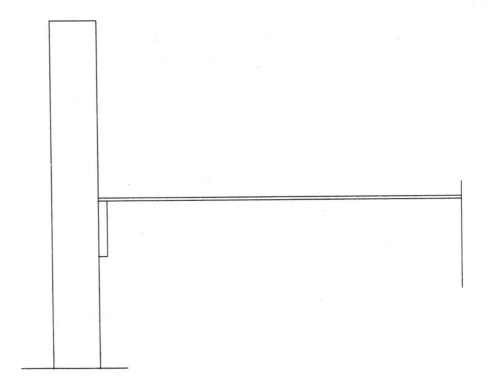

PROBLEM 12-46

Draw a detail of a brick cavity wall supporting 2 x 10 floor joists perpendicular to the wall. Reinforce wall with #4 @ 18" o.c. each way, with metal ties @ ea. 12" vertical @ ec. 32". Use a metal joist connector to each joist.

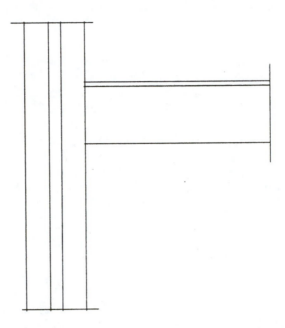

PROBLEM 12-47

Draw the intersection of a scissor truss with a concrete block wall. The roofing is to be 5/12 pitch, cedar shakes. Box in the eave so no tails are exposed. Vent w/2"-wide continuous vent. The wall is to be reinforced with #5 @ 32" o.c. each way. Use a bond beam at the top of the wall with 4-#4 rebar. Solid grout all steel cells (1/2" = 1'-0").

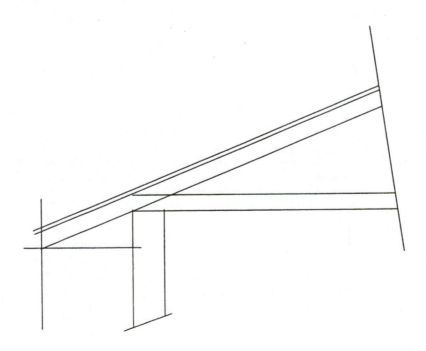

PROBLEM 12-48

Draw the intersection of a 2 x 8 rafter with a concrete clock wall. The roofing is to be 6/12 pitch, covered with built-up roofing. The wall is to be reinforced with #5 @ 32" o.c. each way. Use a bond beam at the top of the wall with 4-#4 rebar. Solid grout all steel cells (1/2" = 1'-0").

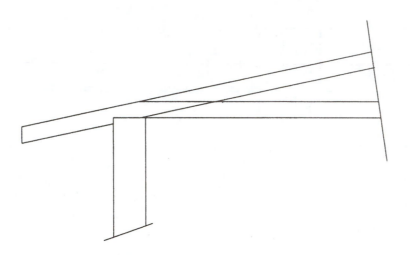

PROBLEM 12-49

Draw the intersection of a truss with a 2 x 6 wall. Trusses to be standard with a 5/12 pitch, cedar shakes. Walls to be double-wall construction with brick veneer. Use a 1/4" x 3" x 3" lintel to support brick. Use 2-2 x 12 hdrs and 2 x 6 nailer. Box in the eave so no tails are exposed. Vent w/2"-wide continuous vent (1" = 1'-0").

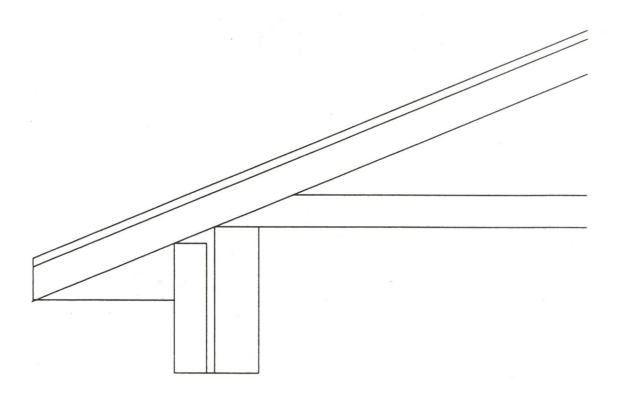

PROBLEM 12-50

Draw the intersection of a 2 x 12 rafter/ceiling joist with a 2 x 6 wall. Use a 7/12 pitch, clay tiles. Walls to be covered with 1" exterior stucco. Determine an overhang so that a 2 x 6 fascia will not interfere with windows at normal height (1/2" = 1'-0").

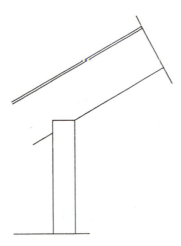

CHAPTER **13**
Sections

DIRECTIONS: Lay out and draw the following sections using required notes and dimensions for materials and sizes typical for your area. Use the floor plan from Chapter 4 and foundation plan from Chapter 11 to determine structural requirements and type of construction. Unless noted, all roof systems will be framed with trusses.

PROBLEM 13-1(A)

Use a scale of 3/8" = 1'-0" to draw a section for the house in Problem 4-1 (page 50).

PROBLEM 13-1(B)

Use a scale of 3/8" = 1'-0" to draw a section for the house in Problem 4-1. Use floor joists for the floor framing system.

PROBLEM 13-1(C)

Use a scale of 3/8" = 1'-0" to draw a section for the house in Problem 4-1. Use a concrete slab for the floor framing system.

PROBLEM 13-2(A)

Using the drawing on page 197 as a guide, draw a full section for the house in Problem 4-2 (page 51). Use floor joists for the floor framing system. Use a scale of 3/8" = 1'-0".

PROBLEM 13-2(B)

Using the partial section on page 197 as a guide, draw a full section for the house in Problem 4-2. Use post-and-beam for the floor framing system. Use a scale of 3/8" = 1'-0".

PROBLEM 13-2(C)

Using the drawing on page 198 as a guide, draw a full section for the house in Problem 4-2. Use a concrete slab for the floor framing system. Use a scale of 3/8" = 1'-0".

PROBLEM 13-1(A)

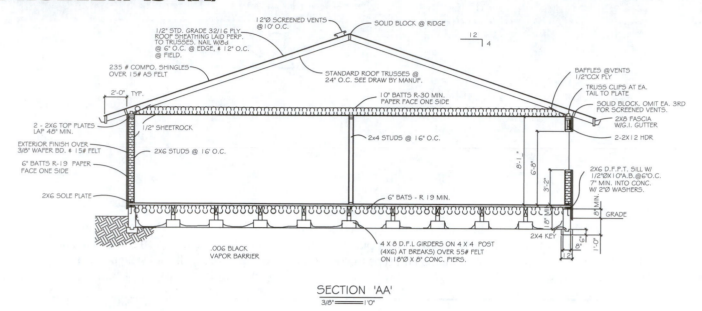

1/2" STD. GRADE 32/16 PLY ROOF SHEATHING LAID PERP. TO TRUSSES. NAIL W/8d @ 6" O.C. @ EDGE, & 12" O.C. @ FIELD.

12"Ø SCREENED VENTS @ 10' O.C.

SOLID BLOCK @ RIDGE

235 # COMPO. SHINGLES OVER 15# AS FELT

12 | 4

STANDARD ROOF TRUSSES @ 24" O.C. SEE DRAW BY MANUF.

BAFFLES @ VENTS 1/2"CCX PLY

2'-0" TYP.

10" BATTS R-30 MIN. PAPER FACE ONE SIDE

TRUSS CLIPS AT EA. TAIL TO PLATE

2 - 2X6 TOP PLATES LAP 48" MIN.

1/2" SHEETROCK

SOLID BLOCK. OMIT EA. 3RD FOR SCREENED VENTS.

2X8 FASCIA W/G.I. GUTTER

EXTERIOR FINISH OVER 3/8" WAFER BD. & 15# FELT

2X6 STUDS @ 16' O.C.

2x4 STUDS @ 16" O.C.

2-2X12 HDR

6" BATTS R-19 PAPER FACE ONE SIDE

8'-1"

6'-8"

2X6 D.F.P.T. SILL W/ 1/2"ØX10"A.B.@6"O.C. 7" MIN. INTO CONC. W/ 2"Ø WASHERS.

2X6 SOLE PLATE

6" BATS - R 19 MIN.

3'-2"

8" MIN

GRADE

18" MIN

2X4 KEY

.006 BLACK VAPOR BARRIER

4 X 8 D.F.L GIRDERS ON 4 X 4 POST (4X6) AT BREAKS) OVER 55# FELT ON 18"Ø X 8" CONC. PIERS.

6"

8"

1'-0"

SECTION 'AA'
3/8" = 1'0"

PROBLEM 13-1(B)

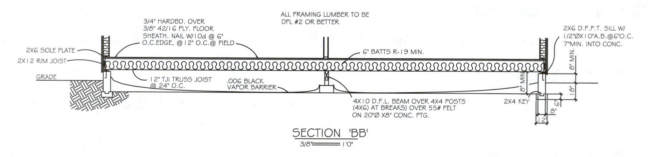

3/4" HARDBD. OVER 3/8" 42/16 PLY. FLOOR SHEATH. NAIL W/10d @ 6" O.C.EDGE, @ 12" O.C.@ FIELD

ALL FRAMING LUMBER TO BE DFL #2 OR BETTER

2X6 D.F.P.T. SILL W/ 1/2"ØX10"A.B.@6"O.C. 7"MIN. INTO CONC.

2X6 SOLE PLATE

2X12 RIM JOIST

6" BATTS R-19 MIN.

8" MIN

GRADE

12" TJI TRUSS JOIST @ 24" O.C.

.006 BLACK VAPOR BARRIER

8" MIN

4X10 D.F.L. BEAM OVER 4X4 POSTS (4X6) AT BREAKS) OVER 55# FELT ON 20"Ø X8" CONC. FTG.

2X4 KEY

18"

6"

8"

12"

SECTION 'BB'
3/8" = 1'0"

PROBLEM 13-1(C)

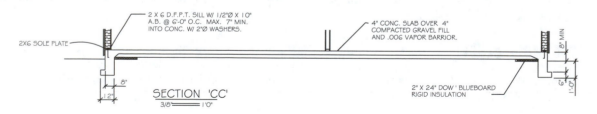

2 X 6 D.F.P.T. SILL W/ 1/2"Ø X 10" A.B. @ 6'-0" O.C. MAX. 7" MIN. INTO CONC. W/ 2"Ø WASHERS.

4" CONC. SLAB OVER 4" COMPACTED GRAVEL FILL AND .006 VAPOR BARRIOR.

2X6 SOLE PLATE

8" MIN

8"

12"

2" X 24" DOW ' BLUEBOARD RIGID INSULATION

6"

1'-0"

SECTION 'CC'
3/8" = 1'0"

PROBLEM 13-2(A)

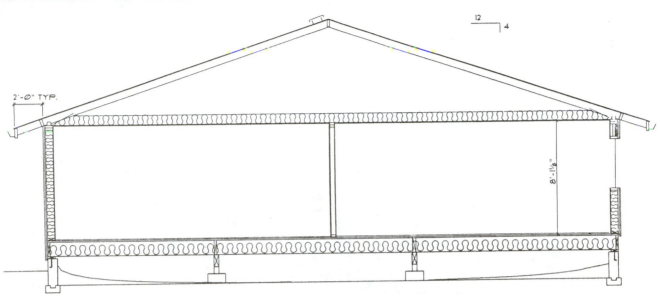

2'-0" TYP.

12
4

8'-1⅛"

PROBLEM 13-2(B)

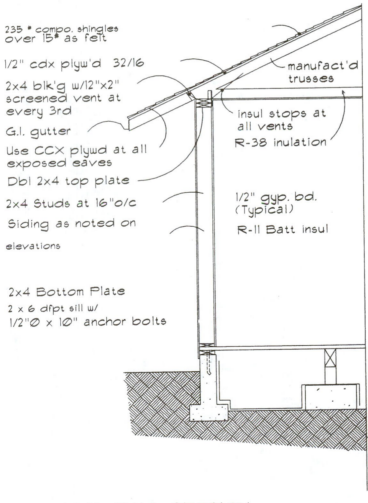

235 # compo. shingles
over 15# as felt

1/2" cdx plyw'd 32/16

2x4 blk'g w/12"x2"
screened vent at
every 3rd

G.I. gutter

Use CCX plywd at all
exposed eaves

Dbl 2x4 top plate

2x4 Studs at 16"o/c

Siding as noted on

elevations

2x4 Bottom Plate
2 x 6 dfpt sill w/
1/2"Ø x 10" anchor bolts

manufact'd
trusses

insul stops at
all vents

R-38 inulation

1/2" gyp. bd.
(Typical)

R-11 Batt insul

TYPICAL WALL SECTION

SCALE 1 1/2" = 1'-0"

PROBLEM 13-2(C)

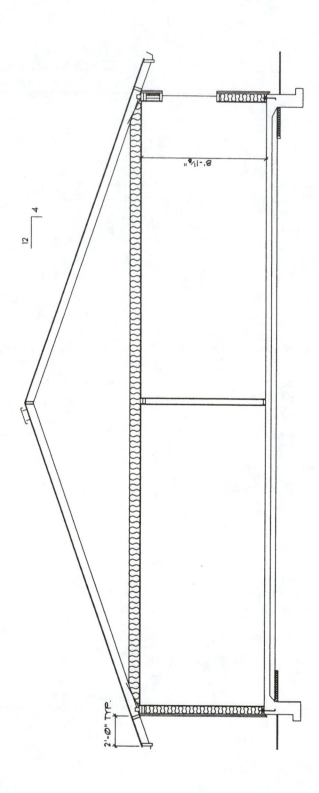

PROBLEM 13-3(A)

1. Using the partial section below as a guide, draw a partial section for the residence in Problem 4-3 (page 52), using proper line quality. Use post-and-beam for the floor framing system. Completely dimension and label per specifications. Use a scale of 3/4" = 1'-0".

2. Using the partial section below as a guide, draw a full section for the house in Problem 4-3. Use post and beam for the floor framing system. Use a scale of 3/8" = 1'-0".

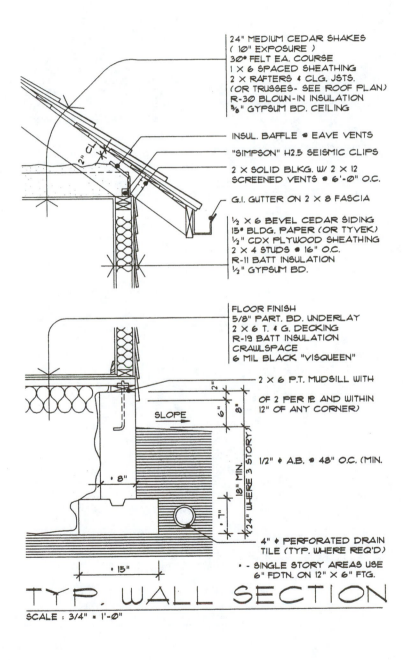

24" MEDIUM CEDAR SHAKES
(10" EXPOSURE)
30# FELT EA. COURSE
1 X 6 SPACED SHEATHING
2 X RAFTERS & CLG. JSTS.
(OR TRUSSES- SEE ROOF PLAN)
R-30 BLOWN-IN INSULATION
⅜" GYPSUM BD. CEILING

INSUL. BAFFLE @ EAVE VENTS

"SIMPSON" H2.5 SEISMIC CLIPS

2 X SOLID BLKG. W/ 2 X 12
SCREENED VENTS @ 6'-0" O.C.

G.I. GUTTER ON 2 X 8 FASCIA

½ X 6 BEVEL CEDAR SIDING
15# BLDG. PAPER (OR TYVEK)
½" CDX PLYWOOD SHEATHING
2 X 4 STUDS @ 16" O.C.
R-11 BATT INSULATION
½" GYPSUM BD.

FLOOR FINISH
5/8" PART. BD. UNDERLAY
2 X 6 T. & G. DECKING
R-19 BATT INSULATION
CRAWLSPACE
6 MIL BLACK "VISQUEEN"

2 X 6 P.T. MUDSILL WITH
OF 2 PER 12 AND WITHIN
12" OF ANY CORNER)

1/2" ø A.B. @ 48" O.C. (MIN.

4" ø PERFORATED DRAIN
TILE (TYP. WHERE REQ'D)

● - SINGLE STORY AREAS USE
6" FDTN. ON 12" X 6" FTG.

SLOPE

TYP. WALL SECTION
SCALE : 3/4" = 1'-0"

PROBLEM 13-3(B)

1. Using the partial section below as a guide, draw a partial section for the residence in Problem 4-3 using proper line quality. Use floor joists for the floor framing system. Completely dimension and label per specifications. Use a scale of 3/4" = 1'-0".

2. Using the partial section below as a guide, draw a full section for the house in Problem 4-3. Use floor joists for the floor framing system and a scale of 3/8" = 1'-0".

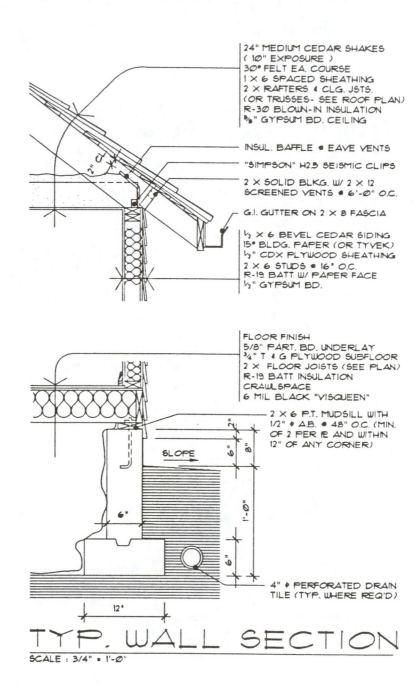

24" MEDIUM CEDAR SHAKES
(10" EXPOSURE)
30# FELT EA. COURSE
1 X 6 SPACED SHEATHING
2 X RAFTERS & CLG. JSTS.
(OR TRUSSES- SEE ROOF PLAN)
R-30 BLOWN-IN INSULATION
⅝" GYPSUM BD. CEILING

INSUL. BAFFLE @ EAVE VENTS

"SIMPSON" H2.5 SEISMIC CLIPS

2 X SOLID BLKG. W/ 2 X 12
SCREENED VENTS @ 6'-0" O.C.

G.I. GUTTER ON 2 X 8 FASCIA

½ X 6 BEVEL CEDAR SIDING
15# BLDG. PAPER (OR TYVEK)
½" CDX PLYWOOD SHEATHING
2 X 6 STUDS @ 16" O.C.
R-19 BATT W/ PAPER FACE
½" GYPSUM BD.

FLOOR FINISH
5/8" PART. BD. UNDERLAY
¾" T & G PLYWOOD SUBFLOOR
2 X FLOOR JOISTS (SEE PLAN)
R-19 BATT INSULATION
CRAWLSPACE
6 MIL BLACK "VISQUEEN"

2 X 6 P.T. MUDSILL WITH
1/2" Ø A.B. @ 48" O.C. (MIN.
OF 2 PER PC. AND WITHIN
12" OF ANY CORNER)

SLOPE

4" Ø PERFORATED DRAIN
TILE (TYP. WHERE REQ'D)

2"
6"
8"
1'-0"
6"
6"
12"

TYP. WALL SECTION
SCALE : 3/4" = 1'-0"

PROBLEM 13-3(C)

1. Using the partial section below as a guide, draw a partial section for the residence in Problem 4-3 using proper line quality. Use a concrete slab for the floor framing system. Completely dimension and label appropriately for your area. Use a scale of 3/4" = 1'-0".

2. Using the partial section below as a guide, draw a full section for the house in Problem 4-3. Use a concrete slab for the floor framing system. Completely label for your area. Use a scale of 3/8" = 1'-0".

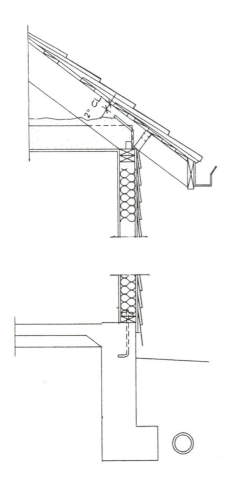

TYP. WALL SECTION

SCALE : 3/4" = 1'-0"

Use the residence in Problem 4–4 (on page 53), and the detail, partial section, and sections below and on pages 203 and 204 as a guide to complete the following problems.

PROBLEM 13-4(A)

Use a scale of 1/4" = 1'-0" to draw, label, and completely dimension a section through the porch, entry, and living room.

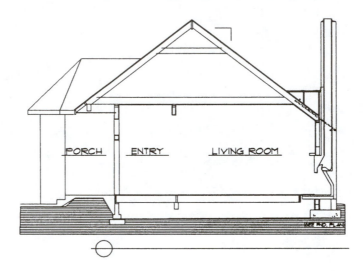

PROBLEM 13-4(B)

Use a scale of 1/4" = 1'-0" to draw, label, and dimension a longitudinal section through the family room and extend it to the master bedroom and wardrobe.

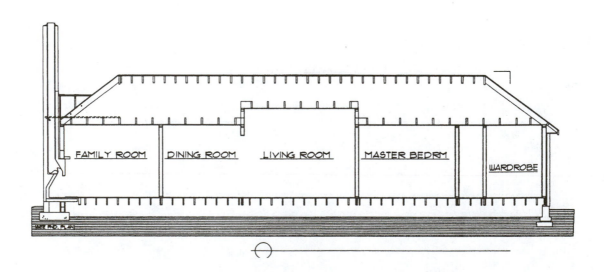

PROBLEM 13-4(C)

Use a scale of 1/4" = 1'-0" to draw, dimension, and label a section through the garage.

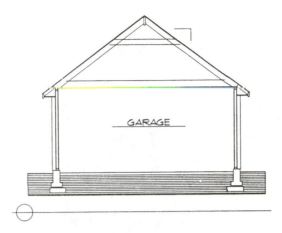

GARAGE

PROBLEM 13-4(D)

Use a scale of 3/4" = 1'-0" to draw, label, and completely dimension a partial section showing typical construction.

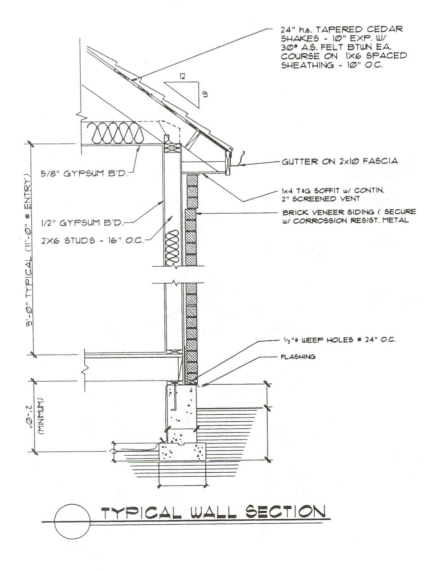

24" h.s. TAPERED CEDAR
SHAKES - 10" EXP. W/
30# A.S. FELT BTWN EA.
COURSE ON 1X6 SPACED
SHEATHING - 10" O.C.

12
9

GUTTER ON 2x10 FASCIA

1x4 T&G SOFFIT w/ CONTIN.
2" SCREENED VENT

BRICK VENEER SIDING (SECURE
w/ CORROSSION RESIST. METAL

5/8" GYPSUM B'D.

1/2" GYPSUM B'D.
2X6 STUDS - 16" O.C.

9'-0" TYPICAL (11'-0" @ ENTRY)

½"∅ WEEP HOLES @ 24" O.C.

FLASHING

2'-0" (MINIMUM)

TYPICAL WALL SECTION

PROBLEM 13-4(E)

Use a scale of 1" = 1'-0" to draw, label, and completely dimension a detail showing support for brick veneer.

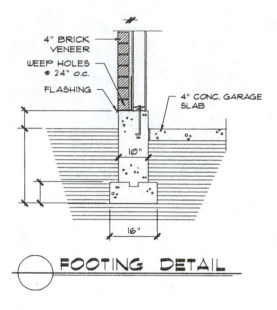

4" BRICK VENEER

WEEP HOLES @ 24" O.C.

FLASHING

4" CONC. GARAGE SLAB

10"

16"

FOOTING DETAIL

Use the residence in Problem 4–4, and the sections on the following pages as a guide to complete the following problems.

PROBLEM 13-5(A)

Use a scale of 3/8" = 1'-0" to draw, label, and dimension a section through the family room and bedroom 3/den.

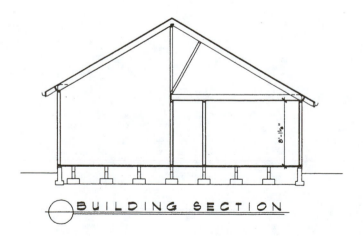

8'-1½"

BUILDING SECTION

PROBLEM 13-5(B)

Use a scale of 1/4" = 1'-0" to draw, label, and dimension a section through patio and sun room.

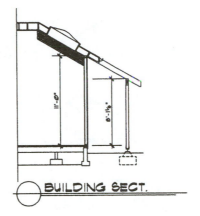

PROBLEM 13-5(C)

Use a scale of 1/4" = 1'-0" to draw, label and dimension a section through the garage.

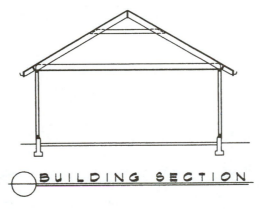

PROBLEM 13-5(D)

Use a scale of 1/4" = 1'-0" to draw, label, and dimension a section through the living room.

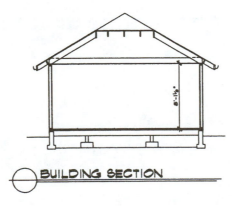

PROBLEM 13-5(E)

Use a scale of 1/4" = 1'-0" to draw, label, and dimension a section through the entry.

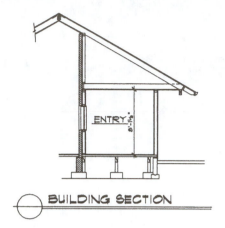

ENTRY

BUILDING SECTION

PROBLEM 13-6(A)

Using a section as a guide, draw a full section for the house in Problem 4-6. Use floor joists for the floor framing system and a scale of 3/8" = 1'-0".

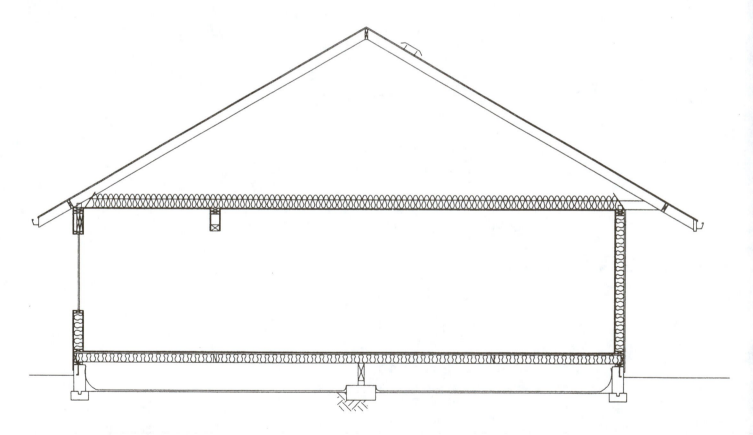

PROBLEM 13-6(B)

Use a scale of 3/8" to draw, label, and dimension a section using post-and-beam framing methods for the floor.

PROBLEM 13-6(C)

Use a scale of 3/8" = 1'-0" to draw, label, and dimension a section using a concrete slab.

Use the residence in Problem 4–7 and the sections on the following pages as a guide to complete the following problems. Chooses a scale appropriate for the information and detail to be drawn.

PROBLEM 13-7(A)

Draw, label, and dimension a section through the garage and courtyard.

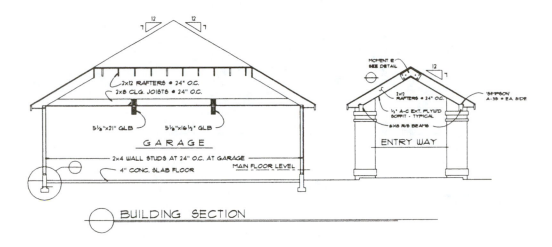

PROBLEM 13-7(B)

Draw, label, and dimension a section through the dining and living rooms.

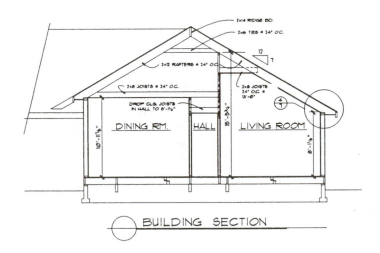

PROBLEM 13-7(C)

Draw, label, and dimension a section through the entry and wardrobe area of the master bedroom. Show the courtyard in the background.

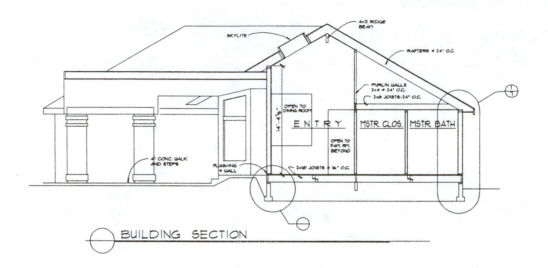

PROBLEM 13-7(D)

Draw, label, and dimension a section through the garage and family room.

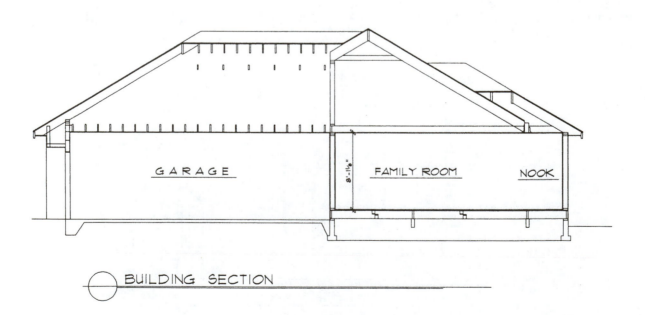

PROBLEM 13-7(E)

Draw, label, and dimension a longitudinal section from the family room to bedroom 2.

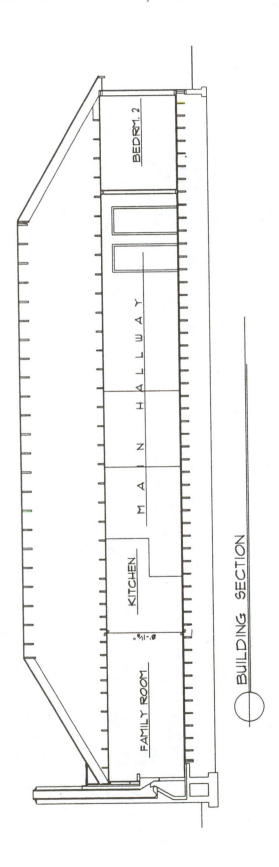

PROBLEM 13-7(F)

Using a scale of 1 1/2" = 1'-0", draw a detail of the rafter connection above the entryway. Specify 3 1/2" dia. bolts thru each rafter 1 1/2" in from the edge of a 1/2" thick x 7" steel plate. Bolts to be 4" o.c. each way. Provide a steel plate centered on each side of each rafter.

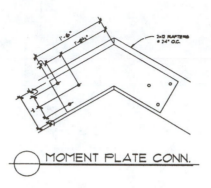

MOMENT PLATE CONN.

PROBLEM 13-7(G)

Draw, label, and dimension a partial section showing typical construction of the entire residence.

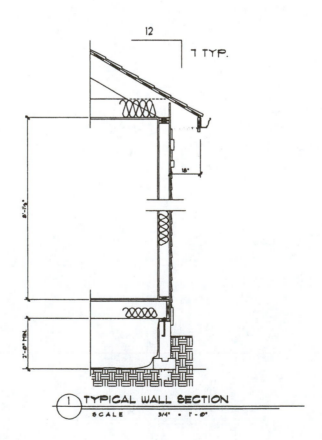

TYPICAL WALL SECTION

SCALE   3/4" = 1'-0"

PROBLEM 13-7(H)

Draw a cornice detail at a scale of 3/4" = 1'-0".

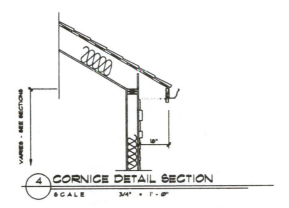

PROBLEM 13-7(I)

Using a scale of 1 1/2" = 1'-0", draw a rake detail.

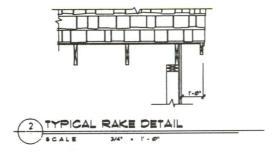

PROBLEM 13-7(J)

Using a scale of 3/4" = 1'-0", draw a detail showing the intersection of the stem wall and the garage slab.

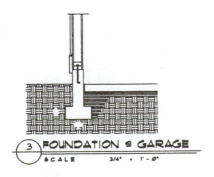

PROBLEM 13-7(K)

Using a scale of 3/4" = 1'-0", draw a detail showing a typical footing with wood floor intersecting a concrete slab.

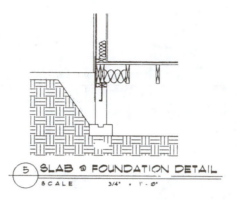

5 SLAB @ FOUNDATION DETAIL
SCALE   3/4" . 1'-0"

PROBLEM 13-8

Using the drawing as a guide, lay out all required sections and details needed to fully describe the residence in Problem 4-8.

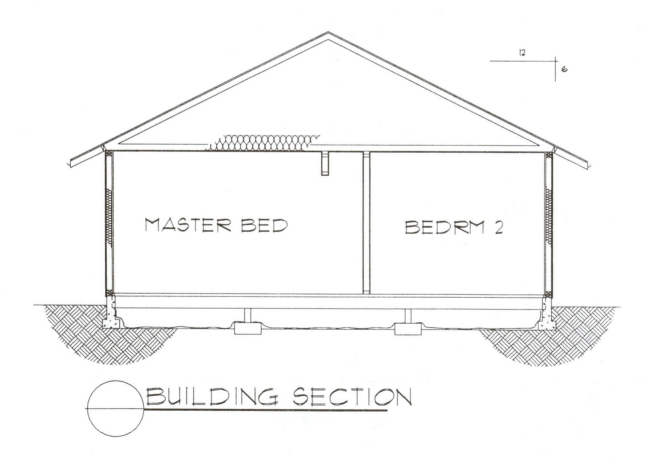

12
6

MASTER BED

BEDRM 2

BUILDING SECTION

PROBLEM 13-9

Using the drawings as a guide, lay out all required sections and details needed to fully describe the residence in Problem 4-9.

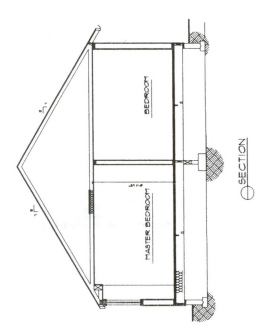

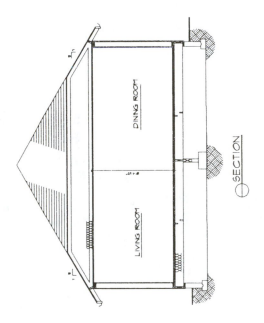

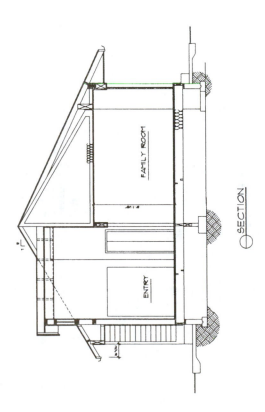

PROBLEM 13-10

Use the floor plan from Chapter 4, the roof plan from Chapter 8, the foundation plan from Chapter 11, the attached preliminary design, and the guidelines in your text to layout, draw, and label a section for the home in Problem 4-10.

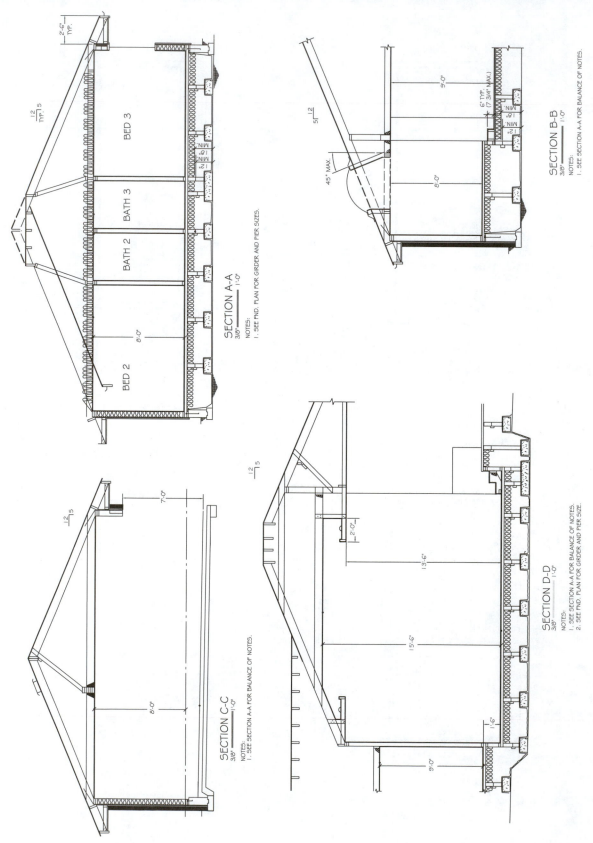

PROBLEM 13-11

Use the floor plan from Chapter 4, the roof plan from Chapter 8, the foundation plan from Chapter 11, the attached preliminary design, and the guidelines in your text to layout, draw, and label a section for the home in Problem 4-11.

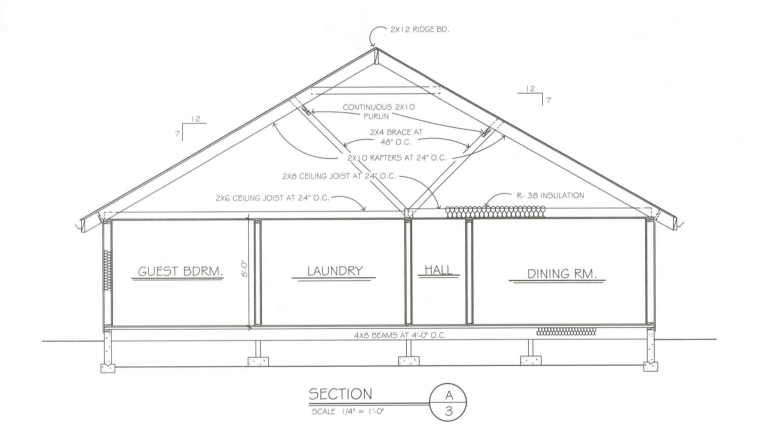

2X12 RIDGE BD.

CONTINUOUS 2X10
PURLIN

2X4 BRACE AT
48" O.C.

2X10 RAFTERS AT 24" O.C.

2X8 CEILING JOIST AT 24" O.C.

2X6 CEILING JOIST AT 24" O.C.

R- 38 INSULATION

12
7

12
7

GUEST BDRM.

LAUNDRY

HALL

DINING RM.

8'-0"

4X8 BEAMS AT 4'-0" O.C.

SECTION
SCALE 1/4" = 1'-0"

A
3

PROBLEM 13-12

Use the floor plan from Chapter 4, the roof plan from Chapter 8, the foundation plan from Chapter 11, the attached preliminary design, and the guidelines in your text to layout, draw, and label a section for the home in Problem 4-12.

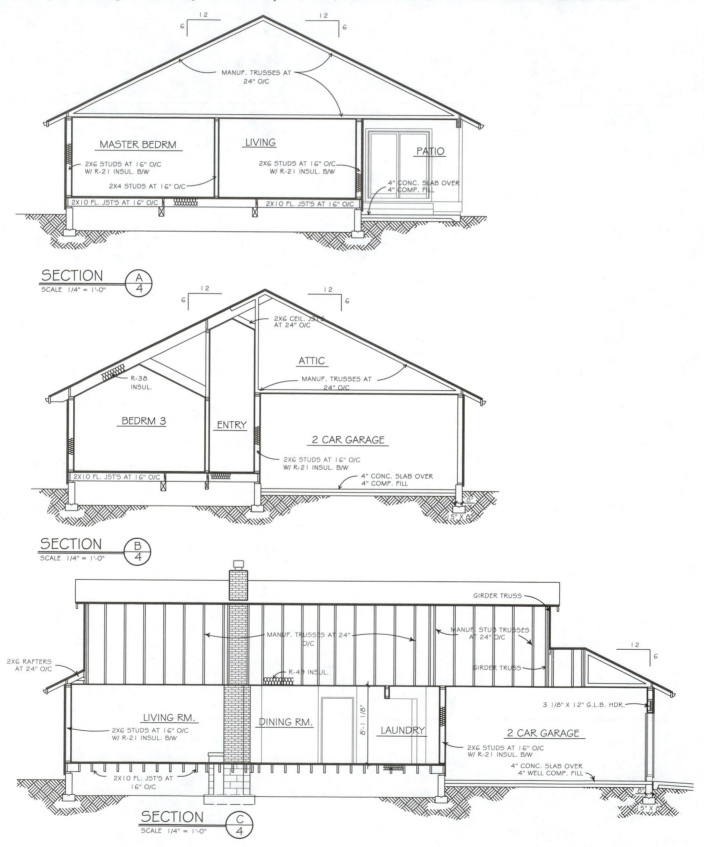

PROBLEM 13-13

Use the floor plan from Chapter 4, the roof plan from Chapter 8, the foundation plan from Chapter 11, the attached preliminary design, and the guidelines in your text to layout, draw, and label a section for the home in Problem 4-13.

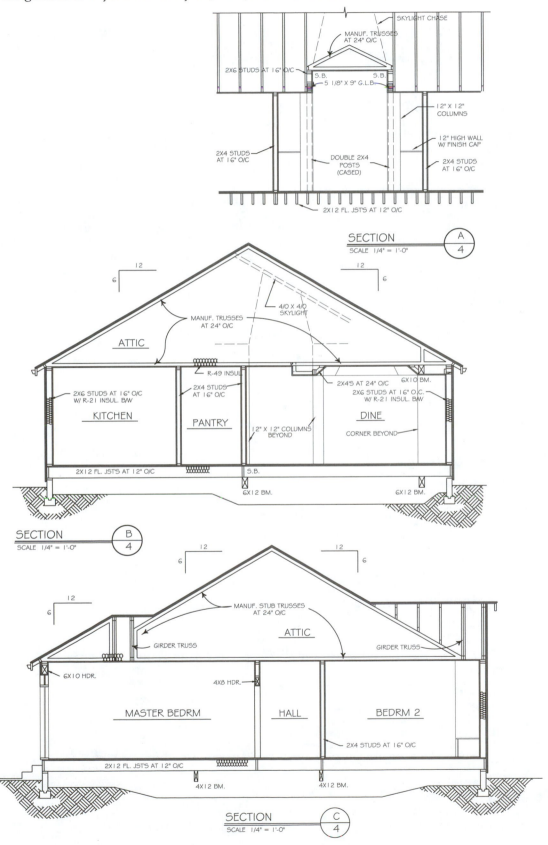

PROBLEM 13-14

Use the floor plan from Chapter 4, the roof plan from Chapter 8, the foundation plan from Chapter 11, the attached preliminary design, and the guidelines in your text to layout, draw, and label a section for the home in Problem 4-14.

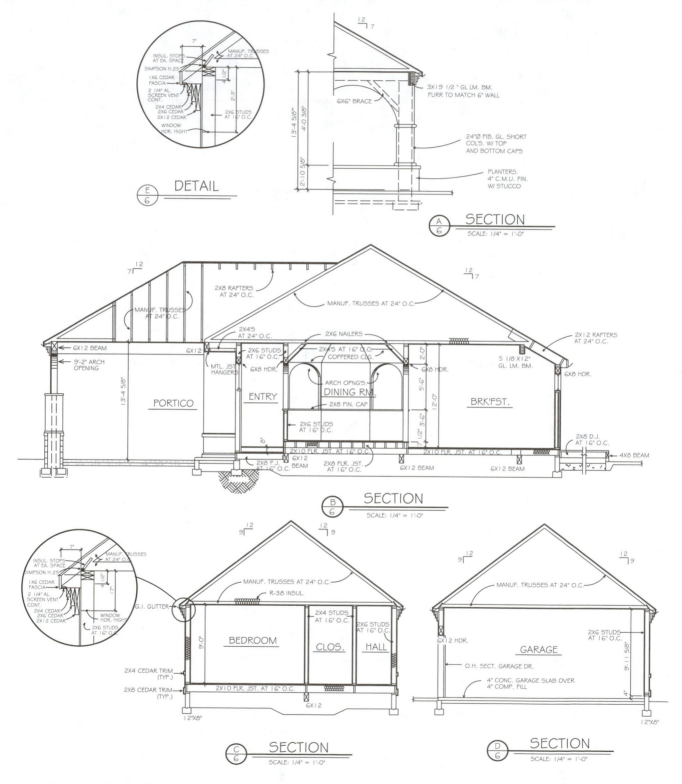

PROBLEM 13-15

Draw all required sections and details needed to describe the residence in Problem 4-10.

CHAPTER 14
STAIR SECTIONS

PROBLEM 14-1

Using the drawing below, list the minimum size of each item.

1. _____

2. _____

3. _____

4. _____

5. _____

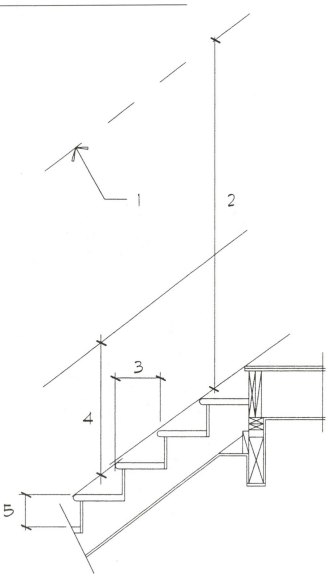

PROBLEM 14-2

Using the drawing below, identify the following stair terms and their typical sizes.

1. _____

2. _____

3. _____

4. _____

5. _____

6. _____

7. _____

8. _____

9. _____

10. _____

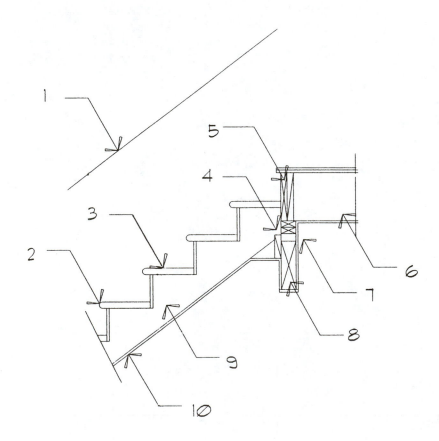

DIRECTIONS: Use a scale of 1/2" = 1'-0" for the following problems. Do your drawings on 8 1/2 x 11" vellum.

PROBLEM 14-3

Design a stairway to connect a tri-level home. The lower floor is 3'-6" below the main floor. The total ceiling height of the lower floor is 8'-0". Calculate the necessary risers based on 7.75" maximum rise and 10 1/2" tread. The lower and middle floors are each 4" concrete slabs. The upper floor is framed with 2 x 8 floor joists. Section the lower flight and show railing at the upper floor.

PROBLEM 14-4

Draw a stairway using 7.5" maximum risers to connect two levels framed with 88 5/8" studs. Use 4 x 12 treads wrapped with carpet. Use 4 x 12 stringers. Specify appropriate metal angles to connect the stringer to floor-framing members and the treads to the stringer.

PROBLEM 14-5

Draw the stairs needed to connect each level of a two-story residence with a basement. The owner would like to have doors at the top and bottom of the basement stairs to help cut the heating bill. The basement ceiling height is 7'-6". The main floor level has 9'-0" ceilings. Each floor is framed with 2 x 10 floor joists. Use oak steps and railing.

PROBLEM 14-6

A stair is needed to connect a deck to a concrete porch. The deck is 4'-2 3/4" above the concrete. The existing deck is framed with 2 x 8 floor joists with 2 x 4 cedar decking laid flat with 1/4" gaps between. The client would like the new stairs to match the existing deck. The owner's only other request is a maximum of 7" risers with 12" wide treads.

PROBLEM 14-7

Design a "U shaped" stair for a one-story residence with a basement. The basement has 8'-0" ceilings. The floor is framed with 2 x 10 joists. The landing is to be 42"-wide minimum and should abut the exterior walls. The upper landing will require a longer total run for the upper flight of steps. Furr the 8" concrete basement wall with 1 x 2's. Frame as required to keep the upper wood walls flush with the lower walls. Treads to be 10 1/2" with as few risers as permitted by the code in your area.

PROBLEM 14-8

Draw a stairway between two wood floors with an elevation difference of 5'-4". Use 10 1/2" treads.

PROBLEM 14-9

Draw a stair to connect two levels that are a total of 9'-6" apart. Use 4x open treads.

PROBLEM 14-10

Design and draw a stair section for the residence that was drawn for Problem 4-15.

CHAPTER
Fireplace Sections

PROBLEM 15-1

Identify the terms in the drawing below.

1. _____
2. _____
3. _____
4. _____
5. _____
6. _____
7. _____

8. _____
9. _____
10. _____
11. _____
12. _____
13. _____
14. _____

15. _____
16. _____
17. _____
18. _____
19. _____
20. _____

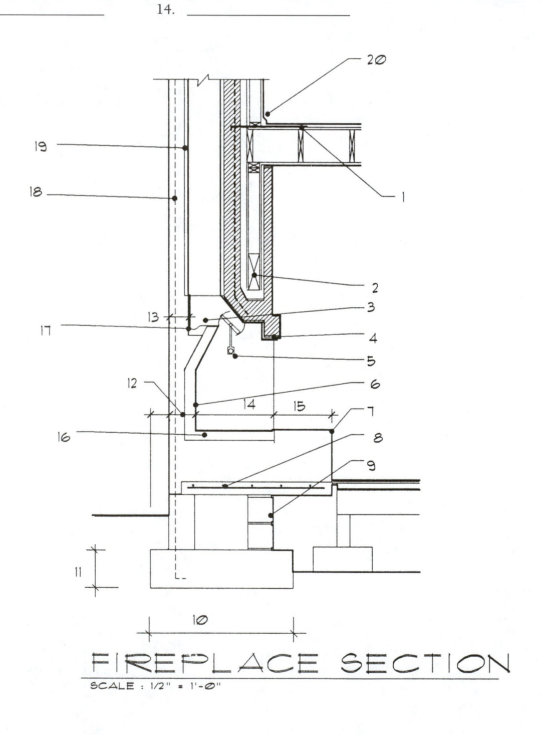

FIREPLACE SECTION
SCALE : 1/2" = 1'-0"

PROBLEM 15-2

Using a scale of 3/8" = 1'-0", draw a section of a masonry fireplace in a house with a concrete floor and truss roof. Use a 14" raised hearth.

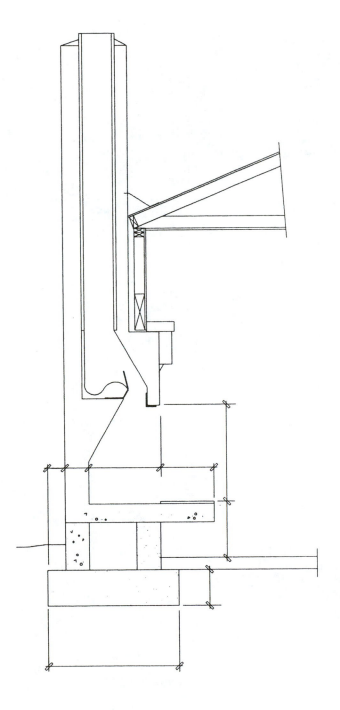

PROBLEM 15-3

Using a scale of 3/8" = 1'-0", draw a section of a masonry fireplace in a house with 2 x 6 floor joists at 24" o.c. and 2 x 12 raft./c.j. with a plywood cant strip.

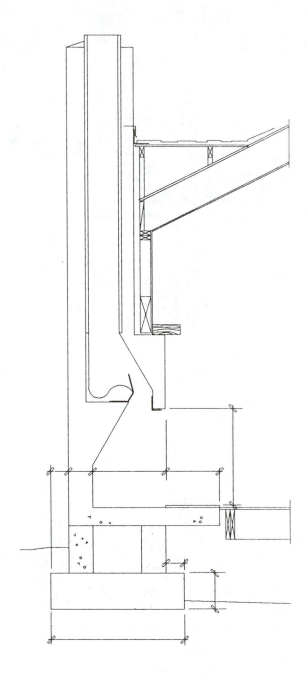

PROBLEM 15-4

Use a scale of 3/8" = 1'-0" to draw a section of a masonry fireplace in a house with 2 x 10 floor joists at 24" o.c. and truss roof. Show all framing members running parallel to the fireplace and chimney. Show a flush hearth and a brick mantel.

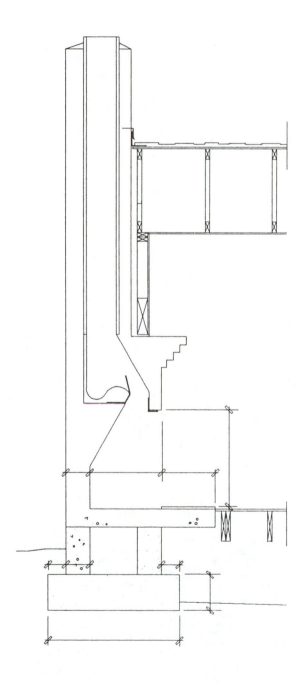

PROBLEM 15-5

Using a scale of 3/8" = 1'-0", draw a section of a concrete-block fireplace and chimney with an exterior face of stucco, in a house with a concrete floor and truss roof.

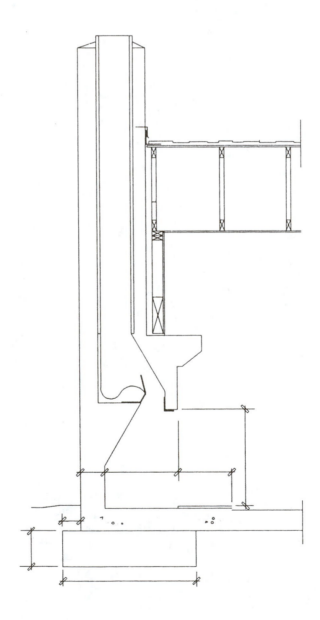

DIRECTIONS: Use a scale of 3/4" = 1'-0" for the following problems unless noted. Use "generic lettering" on stock details (floor joist—see framing plan) to cover a wide variety of uses.

PROBLEM 15-6

Use the drawing below to show and specify all reinforcing for the following chimney. Specify #4 rebars @ 12" o.c. each way in the foundation.

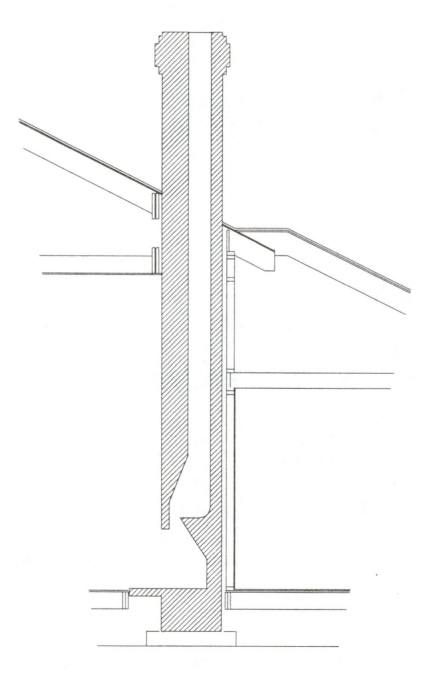

PROBLEM 15-7

Using the drawing below as a guide, draw three different elevations without changing the size of the firebox or the hearth height.

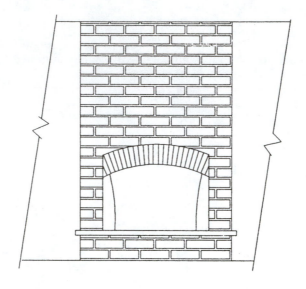

PROBLEM 15-8

Design and draw a masonry fireplace elevation for a chimney. The lower floor is a concrete slab. Plate height is 10', and the roof pitch is 3/12. The face of the fireplace is parallel to the 2 x 12 rafters/c.j. The chimney is five feet wide, and the center of the chimney is six feet from the 10'-high wall. Design an elevation that will accent the height. The owners would like a "massive oak mantel" but are unable to clearly express their desires. Design two different options that combine many materials.

PROBLEM 15-9

Design and draw a stock detail to be used for a two-story fireplace and chimney using masonry. Each ceiling is eight feet high. The bottom floor is a concrete slab, the upper floor is framed with 2 x 10 floor joists, and the ceiling is framed with 2 x 6 ceiling joists and rafters.

PROBLEM 15-10

Design and draw a fireplace section and elevation for the floor plan that was drawn in Chapter 4.

Design Criteria and Structural Loading

DIRECTIONS: Complete the following problems on a separate sheet of paper. Draw a loading sketch for each problem and show all math work.

PROBLEM 16-1

If a joist with a total uniform load of 500 lbs is supported at each end, how much weight will each end be supporting?

PROBLEM 16-2

If a joist is 14' long with a total uniform load of 700 lbs is supported at the midpoint and at each end, how much weight will be supported at each support?

PROBLEM 16-3

A beam is 20' long and supports a 15'-wide area of tile roof. What weight will the support post at each end be supporting?

PROBLEM 16-4

A 16 foot rafter ceiling joist at a 6/12 pitch will be supporting a cedar shake roof. What will be the total weight supported if the rafters are at 12" spacings? 16"? 24"?

PROBLEM 16-5

A truss with 24" overhangs will span 28' over a residence. It will be supported by a wall at one end and a girder truss with a metal hanger at the other end. How much weight will the hanger need to support?

PROBLEM 16-6

Steel columns will be used to support a 16'-long girder with a weight of 1,260 lbs per linear foot. How much weight will the columns support?

PROBLEM 16-7

A 24'-wide residence will be built with a truss roof, with built-up roofing and 30" overhangs. How much weight would a header over an 8'-wide window need to support?

DIRECTIONS: Use the following information to complete Problems 16–8 through 16–12. Answer the questions by providing the weight per linear foot. Provide a sketch for each problem to represent the loads.

A one-level residence will be built using trusses with 36" overhangs. The residence will be 24' wide. The roof is 235 lb composition shingles.

PROBLEM 16-8

How much weight will be supported by a header over a 6'-wide window in an exterior wall?

PROBLEM 16-9

A wall is to be built 11'-9" from the right bearing wall. If a 2'6" pocket door is to be installed in the wall, how long will the header need to be and how much weight will it be supporting?

PROBLEM 16-10

Determine the load to be supported and specify the width of footing required if a wall is placed 14' from the left bearing wall.

PROBLEM 16-11

What will be the total dead load that the bearing wall on the right side will be supporting?

PROBLEM 16-12

What will be the total load that the bearing wall on the left side will be supporting?

DIRECTIONS: Use the following information to complete Problems 16–13 through 16-17. Answer the question by providing the weight per linear foot and total weight. Provide a sketch for each problem to represent the floor plan and the load to be supported.

A 26'-wide residence will be built with a main floor over a daylight basement. The roof will be framed with trusses with a 30" overhang. The roofing will be clay tiles. The main floor will be supported by a steel beam 14' from the left side. The left side of the basement will be a concrete retaining wall, and the right side will be a wood wall. The residence is 32' long.

PROBLEM 16-13

A wall will be on the main floor 14'-6" from the left exterior wall. How much weight will be supported?

PROBLEM 16-14

The steel beam will be supported at 10' from each end wall by 3" dia. steel columns. How much weight will each column be supporting?

PROBLEM 16-15

How much weight will be supported at each exterior end of the beam described in the last question?

PROBLEM 16-16

How much dead load will be supported by the footing on the right side of the basement?

PROBLEM 16-17

If the retaining wall is 8' high, how much pressure must be resisted per linear foot of wall if the soil P.S.I. is 25 P.S.F.?

DIRECTIONS: Use the following information to complete Problems 16–18 through 16–25. Answer the questions by providing the weight per linear foot and total weight.

A 32' wide, two story residence is to be built with cedar shakes. The roof is to be stick-framed and supported by a bearing wall 15' from the left bearing wall. The overhangs will be 18". The upper floor will be supported by a bearing wall 16' from the left bearing wall. The lower floor will have a girder directly below the wall on the main floor with supports at 4' intervals. The residence is 32' x 28'.

PROBLEM 16-18

How much weight will be supported by the upper bearing wall near the center of the residence?

PROBLEM 16-19

How much weight will be supported by the upper right exterior bearing wall?

PROBLEM 16-20

How much weight will be supported by the bearing wall near the center of the residence on the main floor?

PROBLEM 16-21

How much weight will be supported by the girder below the center wall?

PROBLEM 16-22

How much weight will be supported by the pier at each end of the girder below the center wall?

PROBLEM 16-23

What will be the total amount of weight supported by the upper bearing wall on the right side of the residence?

PROBLEM 16-24

What will be the total amount of weight supported by the lower bearing wall on the right side of the residence?

PROBLEM 16-25

What will be the total dead load supported by the foundation on the right side of the residence?

PROBLEM 16-26

Compare the spans of a 2 x 8 floor joist at standard spacing for each of the four common species of lumber.

PROBLEM 16-27

Compare the spans of a 2 x 6 ceiling joist at standard spacing for each of the four common species of lumber.

PROBLEM 16-28

Compare the size rafter needed at 24" spacing for each of the four common species of lumber, to span 13'-0" @ 5/12 pitch with 235 lb compo, shingles and a 30 lb live load.

DIRECTIONS: Solve the following problems using the correct formula, on a separate sheet of paper. If a CADD program is used, provide a printout of the problem. Show all work and write each formula to be used. Provide a sketch showing a floor plan and section to illustrate how weight is distributed.

PROBLEM 16-29

Determine the minimum size beam needed to span 14' with a concentrated load of 2,200 lbs in the center. What size piers will be needed to resist the reaction if soil loads are 2,500 P.S.F.?

PROBLEM 16-30

Determine what size joists will be needed to cantilever 18" at a deck if a spacing of 16" is used.

PROBLEM 16-31

Using S.P.F., determine what size beam will be needed to span 15'-6" with a point load of 2,750 lbs. at the midpoint. What size piers will be required to support the loads if soil pressure is 2,000 lbs.?

PROBLEM 16-32

Use D.F.L. and determine what size beam will be required to support the loads in the problem 16-31.

PROBLEM 16-33

A 4 x 8 beam, with an L = 8' and a W = 3,200 lbs will be used to span an opening for a window. Will it fail in Fv?

PROBLEM 16-34

A 6 x 12 is being used as a ridgebeam. L= 14.5', W = 5,500 lbs. Will this beam have a safe E value?

PROBLEM 16-35

If the beam in the last question was a glu-lam beam with an fb of 2400, what size would be required?

PROBLEM 16-36

If the soil-bearing pressure is 2,000 P.S.F. and the concrete has a strength of 3,500 P.S.I., what diameter footing is needed to support a load of 5,500 pounds?

PROBLEM 16-37

What size girder will be required if w = 800 and L = 8'?

PROBLEM 16-38

A beam is 13' long, with a load of 2,700 pounds, 36" from the right end. What size beam is required?

PROBLEM 16-39

A 16'-long laminated ridge beam will support 800 pounds per linear foot. Determine what depth 5 1/8 beam should be used.

PROBLEM 16-40

The ridge beam in the last question will be supported at one end by a post and at the other end by a 4'-long beam over a door. The point load from the ridge beam will be 12" from one end of the four-foot header. What size door header should be used?

Construction Specifications, Permits, and Contracts

PROBLEM 17-1

DIRECTIONS: Use the following blank copy of the FHA Description of Materials form. Complete the form by preparing the construction specifications for the house that you have been drawing, starting with one of the floor plan problems from Chapter 4. It is suggested that you copy the FHA Description of Materials form for use as a rough draft and all preliminary work. The final completion of the form should be professionally hand-lettered or typed depending on your course guidelines. In order to complete the form you may have to research local suppliers of building products in an effort to select specific items for your house.

PROBLEM 17-1

VETERANS ADMINISTRATION, U.S.D.A. FARMERS HOME ADMINISTRATION, AND

U.S. DEPARTMENT OF HOUSING AND URBAN DEVELOPMENT
HOUSING - FEDERAL HOUSING COMMISSIONER   OMB Approval No. 2502-0192 (Exp. 6-30-87)
For accurate register of carbon copies, form may be separated along above
fold. Staple completed sheets together in original order.

☐ Proposed Construction   **DESCRIPTION OF MATERIALS**   No. _____
(To be inserted by HUD, VA or FmHA)

☐ Under Construction

Property address _____ City _____ State _____

Mortgagor or Sponsor _____
(Name)   (Address)

Contractor or Builder _____
(Name)   (Address)

INSTRUCTIONS

1. For additional information on how this form is to be submitted, number of copies, etc., see the instructions applicable to the HUD Application for Mortgage Insurance, VA Request for Determination of Reasonable Value, or FmHA Property Information and Appraisal Report, as the case may be.
2. Describe all materials and equipment to be used, whether or not shown on the drawings, by marking an X in each appropriate check-box and entering the information called for each space. If space is inadequate, enter "See misc." and describe under item 27 or on an attached sheet. THE USE OF PAINT CONTAINING MORE THAN THE PERCENTAGE OF LEAD BY WEIGHT PERMITTED BY LAW IS PROHIBITED.
3. Work not specifically described or shown will not be considered unless

required, then the minimum acceptable will be assumed. Work exceeding minimum requirements cannot be considered unless specifically described.
4. Include no alternates, "or equal" phrases, or contradictory items. (Consideration of a request for acceptance of substitute materials or equipment is not thereby precluded.)
5. Include signatures required at the end of this form.
6. The construction shall be completed in compliance with the related drawings and specifications, as amended during processing. The specifications include this Description of Materials and the applicable Minimum Property Standards.

1. EXCAVATION:

Bearing soil, type _____

2. FOUNDATIONS:

Footings: concrete mix _____; strength psi _____ Reinforcing _____
Foundation wall: material _____ Reinforcing _____
Interior foundation wall: material _____ Party foundation wall _____
Columns: material and sizes _____ Piers: material and reinforcing _____
Girders: material and sizes _____ Sills: material _____
Basement entrance areaway _____ Window areaways _____
Waterproofing _____ Footing drains _____
Termite protection _____
Basementless space: ground cover _____; insulation _____; foundation vents _____
Special foundations _____
Additional information: _____

3. CHIMNEYS:

Material _____ Prefabricated *(make and size)* _____
Flue lining: material _____ Heater flue size _____ Fireplace flue size _____
Vents *(material and size)*: gas or oil heater _____; water heater _____
Additional information: _____

4. FIREPLACES:

Type: ☐ solid fuel; ☐ gas-burning; ☐ circulator *(make and size)* _____ Ash dump and clean-out _____
Fireplace: facing _____; lining _____; hearth _____; mantel _____
Additional information: _____

5. EXTERIOR WALLS:

Wood frame: wood grade, and species _____ ☐ Corner bracing. Building paper or felt _____
Sheathing _____; thickness _____; width _____; ☐ solid; ☐ spaced _____" o. c.; ☐ diagonal; _____
Siding _____; grade _____; type _____; size _____; exposure _____"; fastening _____
Shingles _____; grade _____; type _____; size _____; exposure _____"; fastening _____
Stucco _____; thickness _____", Lath _____, weight _____ lb.
Masonry veneer _____ Sills _____ Lintels _____ Base flashing _____
Masonry: ☐ solid ☐ faced ☐ stuccoed; total wall thickness _____"; facing thickness _____"; facing material _____
Backup material _____; thickness _____"; bonding _____
Door sills _____ Window sills _____ Lintels _____ Base flashing _____
Interior surfaces: dampproofing, _____ coats of _____; furring _____
Additional information: _____
Exterior painting: material _____; number of coats _____
Gable wall construction: ☐ same as main walls, ☐ other construction _____

6. FLOOR FRAMING:

Joists: wood, grade, and species _____; other _____; bridging _____; anchors _____
Concrete slab: ☐ basement floor; ☐ first floor; ☐ ground supported; ☐ self-supporting; mix _____; thickness _____",
reinforcing _____; insulation _____; membrane _____
Fill under slab: material _____; thickness _____". Additional information: _____

7. SUBFLOORING: *(Describe underflooring for special floors under item 21.)*

Material: grade and species _____; size _____; type _____
Laid: ☐ first floor; ☐ second floor; ☐ attic _____ sq. ft.; ☐ diagonal; ☐ right angles. Additional information: _____

8. FINISH FLOORING: *(Wood only. Describe other finish flooring under item 21.)*

LOCATION	ROOMS	GRADE	SPECIES	THICKNESS	WIDTH	BLDG. PAPER	FINISH
First floor							
Second floor							
Attic floor	sq. ft						

Additional information: _____

DESCRIPTION OF MATERIALS
HUD-92005(10-84) HUD HB 4145.1
VA Form 26-1852 Form FmHA 424-2

PROBLEM 17-1 (CONTINUED)

DESCRIPTION OF MATERIALS

9. PARTITION FRAMING:
Studs: wood, grade, and species _____ size and spacing _____ Other _____
Additional information _____

10. CEILING FRAMING:
Joists: wood, grade, and species _____ Other _____ Bridging _____
Additional information _____

11. ROOF FRAMING:
Rafters: wood, grade, and species _____ Roof trusses (see detail): grade and species _____
Additional information _____

12. ROOFING:
Sheathing: wood, grade, and species _____ ; ☐ solid; ☐ spaced _____ ” o.c.
Roofing _____ ; grade _____ ; size _____ ; type _____
Underlay _____ ; weight or thickness _____ , size _____ , fastening _____
Built-up roofing _____ ; number of plies _____ ; surfacing material _____
Flashing: material _____ ; gage or weight _____ ; ☐ gravel stops; ☐ snow guards
Additional information _____

13. GUTTERS AND DOWNSPOUTS:
Gutters: material _____ ; gage or weight _____ ; size _____ ; shape _____
Downspouts: material _____ ; gage or weight _____ ; size _____ ; shape _____ ; number _____
Downspouts connected to: ☐ Storm sewer; ☐ sanitary sewer; ☐ dry-well. ☐ Splash blocks: material and size _____
Additional information _____

14. LATH AND PLASTER
Lath ☐ walls, ☐ ceilings: material _____ ; weight or thickness _____ Plaster: coats _____ ; finish _____
Dry-wall ☐ walls, ☐ ceilings: material _____ ; thickness _____ ; finish _____ ;
Joint treatment _____

15. DECORATING: *(Paint, wallpaper, etc.)*

Rooms	Wall Finish Material and Application	Ceiling Finish Material and Application
Kitchen _____		
Bath _____		
Other _____		

Additional information: _____

16. INTERIOR DOORS AND TRIM:
Doors: type _____ ; material _____ ; thickness _____
Door trim: type _____ ; material _____ Base: type _____ ; material _____ ; size _____
Finish: doors _____ ; trim _____
Other trim (item, type and location) _____
Additional information _____

17. WINDOWS:
Windows: type _____ ; make _____ ; material _____ ; sash thickness _____
Glass: grade _____ ; ☐ sash weights; ☐ balances, type _____ ; head flashing _____
Trim: type _____ ; material _____ Paint _____ ; number coats _____
Weatherstripping: type _____ ; material _____ Storm sash, number _____
Screens: ☐ full; ☐ half, type _____ ; number _____ ; screen cloth material _____
Basement windows: type _____ ; material _____ ; screens, number _____ ; Storm sash, number _____
Special windows _____
Additional information _____

18. ENTRANCES AND EXTERIOR DETAIL:
Main entrance door: material _____ ; width _____ ; thickness _____ ”. Frame: material _____ , thickness _____ ”
Other entrance doors: material _____ ; width _____ ; thickness _____ ”. Frame: material _____ , thickness _____ ”
Head flashing _____ Weatherstripping: type _____ ; saddles _____
Screen doors: thickness _____ ”; number _____ ; screen cloth material _____ Storm doors: thickness _____ ”, number _____
Combination storm and screen doors: thickness _____ ”; number _____ ; screen cloth material _____
Shutters: ☐ hinged; ☐ fixed. Railings _____ , Attic louvers _____
Exterior millwork: grade and species _____ Paint _____ ; number coats _____
Additional information: _____

19. CABINETS AND INTERIOR DETAIL:
Kitchen cabinets, wall units: material _____ ; lineal feet of shelves _____ ; shelf width _____
Base units: material _____ ; counter top _____ ; edging _____
Back and end splash _____ Finish of cabinets _____ ; number coats _____
Medicine cabinets: make _____ ; model _____
Other cabinets and built-in furniture _____
Additional information: _____

20. STAIRS:

Stair	Treads		Risers		Strings		Handrail		Balusters	
	Material	Thickness	Material	Thickness	Material	Size	Material	Size	Material	Size
Basement _____										
Main _____										
Attic _____										

Disappearing: make and model number _____
Additional information: _____

2

PROBLEM 17-1 (CONTINUED)

21. SPECIAL FLOORS AND WAINSCOT: *(Describe Carpet as listed in Certified Products Directory)*

	LOCATION	MATERIAL, COLOR, BORDER, SIZES, GAGE, ETC.	THRESHOLD MATERIAL	WALL BASE MATERIAL	UNDERFLOOR MATERIAL
FLOORS	Kitchen				
	Bath				

	LOCATION	MATERIAL, COLOR, BORDER, CAP SIZES, GAGE, ETC.	HEIGHT	HEIGHT OVER TUB	HEIGHT IN SHOWERS (FROM FLOOR)
WAINSCOT	Bath				

Bathroom accessories: ☐ Recessed; material _____; number _____. ☐ Attached, material _____, number _____

Additional information: _____

22. PLUMBING:

FIXTURE	NUMBER	LOCATION	MAKE	MFR'S FIXTURE IDENTIFICATION NO	SIZE	COLOR
Sink						
Lavatory						
Water closet						
Bathtub						
Shower over tub △						
Stall shower △						
Laundry trays						

△☐ Curtain rod   △☐ Door   ☐ Shower pan: material _____

Water supply: ☐ public; ☐ community system, ☐ individual (private) system. ★

Sewage disposal: ☐ public; ☐ community system, ☐ individual (private) system. ★

★ Show and describe individual system in complete detail in separate drawings and specifications according to requirements

House drain (inside): ☐ cast iron; ☐ tile, ☐ other _____ House sewer (outside): ☐ cast iron, ☐ tile, ☐ other _____

Water piping: ☐ galvanized steel, ☐ copper tubing, ☐ other _____ Sill cocks, number _____

Domestic water heater: type _____; make and model _____, heating capacity _____

_____ gph. 100° rise. Storage tank: material _____, capacity _____ gallons.

Gas service: ☐ utility company; ☐ liq. pet. gas; ☐ other _____ Gas piping ☐ cooking; ☐ house heating

Footing drains connected to: ☐ storm sewer; ☐ sanitary sewer; ☐ dry well. Sump pump; make and model _____

_____; capacity _____; discharges into _____

23. HEATING:

☐ Hot water. ☐ Steam. ☐ Vapor. ☐ One-pipe system. ☐ Two-pipe system.

☐ Radiators. ☐ Convectors. ☐ Baseboard radiation. Make and model _____

Radiant panel: ☐ floor; ☐ wall; ☐ ceiling. Panel coil: material _____

☐ Circulator. ☐ Return pump. Make and model _____, capacity _____ gpm.

Boiler: make and model _____ Output _____ Btuh, net rating _____ Btuh.

Additional information: _____

Warm air: ☐ Gravity. ☐ Forced. Type of system _____

Duct material: supply _____; return _____ Insulation _____, thickness _____ ☐ Outside air intake.

Furnace: make and model _____ Input _____ Btuh, output _____ Btuh.

Additional information: _____

☐ Space heater; ☐ floor furnace; ☐ wall heater. Input _____ Btuh; output _____ Btuh, number units _____

Make, model _____ Additional information: _____

Controls: make and types _____

Additional information: _____

Fuel: ☐ Coal; ☐ oil; ☐ gas; ☐ liq. pet. gas; ☐ electric; ☐ other _____; storage capacity _____

Additional information: _____

Firing equipment furnished separately: ☐ Gas burner, conversion type. ☐ Stoker: hopper feed ☐, bin feed ☐

Oil burner: ☐ pressure atomizing; ☐ vaporizing _____

Make and model _____ Control _____

Additional information: _____

Electric heating system: type _____ Input _____ watts, @ _____ volts, output _____ Btuh.

Additional information: _____

Ventilating equipment: attic fan, make and model _____; capacity _____ cfm.

kitchen exhaust fan, make and model _____

Other heating, ventilating, or cooling equipment _____

24. ELECTRIC WIRING:

Service: ☐ overhead; ☐ underground. Panel: ☐ fuse box; ☐ circuit-breaker, make _____ AMP's _____ No. circuits _____

Wiring: ☐ conduit; ☐ armored cable; ☐ nonmetallic cable; ☐ knob and tube, ☐ other _____

Special outlets: ☐ range; ☐ water heater; ☐ other _____

☐ Doorbell. ☐ Chimes. Push-button locations _____ Additional information _____

25. LIGHTING FIXTURES:

Total number of fixtures _____ Total allowance for fixtures, typical installation, $ _____

Nontypical installation _____

Additional information _____

PROBLEM 17-1 (CONTINUED)

DESCRIPTION OF MATERIALS

26. INSULATION:

Location	Thickness	Material, Type, and Method of Installation	Vapor Barrier
Roof			
Ceiling			
Wall			
Floor			

27. **MISCELLANEOUS:** (Describe any main dwelling materials, equipment, or construction items not shown elsewhere, or use to provide additional information where the space provided was inadequate. Always reference by item number to correspond to numbering used on this form.) _____

_____

_____

_____

_____

_____

_____

_____

_____

_____

HARDWARE: (make, material, and finish.) _____

SPECIAL EQUIPMENT: (State material or make, model and quantity. Include only equipment and appliances which are acceptable by local law, custom and applicable FHA standards. Do not include items which, by established custom, are supplied by occupant and removed when he vacates premises or chattels prohibited by law from becoming realty.)_____

_____

_____

_____

_____

_____

_____

PORCHES:

_____

_____

TERRACES:

_____

_____

GARAGES:

_____

_____

WALKS AND DRIVEWAYS:

Driveway: width _____ ; base material _____ ; thickness _____"; surfacing material _____ ; thickness _____"
Front walk: width _____ ; material _____ ; thickness _____". Service walk: width _____ ; material _____ ; thickness _____"
Steps: material _____ ; treads _____"; risers _____". Cheek walls _____

OTHER ONSITE IMPROVEMENTS:

(Specify all exterior onsite improvements not described elsewhere, including items such as unusual grading, drainage structures, retaining walls, fence, railings, and accessory structures.)

_____

_____

LANDSCAPING, PLANTING, AND FINISH GRADING:

Topsoil _____" thick: ☐ front yard; ☐ side yards; ☐ rear yard to _____ feet behind main building.
Lawns (seeded, sodded, or sprigged): ☐ front yard _____ ; ☐ side yards _____ ; ☐ rear yard_____
Planting: ☐ as specified and shown on drawings; ☐ as follows:

_____ Shade trees, deciduous, _____" caliper.	_____ Evergreen trees. _____' to _____', B & B.
_____ Low flowering trees, deciduous, _____' to _____'	_____ Evergreen shrubs. _____' to _____', B & B.
_____ High-growing shrubs, deciduous, _____' to _____'	_____ Vines, 2-year _____
_____ Medium-growing shrubs, deciduous, _____' to _____'	_____
_____ Low-growing shrubs, deciduous, _____' to _____'	

IDENTIFICATION.—This exhibit shall be identified by the signature of the builder, or sponsor, and/or the proposed mortgagor if the latter is known at the time of application.

Date_____   Signature _____

Signature _____

DIRECTIONS: All forms must be neatly hand-lettered or typed, unless otherwise specified by your instructor.

PROBLEM 17-2

Complete the building permit application on page 235 based on this information.

- Project location address: 3456 Barrington Drive, your city and state
- Nearest cross street: Washington Street
- Subdivision name: Barrington Heights
- Lot 8, Block 2
- Township: 2S
- Range: 1E
- Section: 36
- Tax lot: 2400
- Lot size: 15000 sq ft.
- Building area: 2000 sq ft.
- Basement area: None
- Garage area: 576 sq ft.
- Stories: 1
- Bedrooms: 3
- Water source: Public
- Sewage disposal: Public
- Estimated cost of labor and materials: $88,500
- Plans and specifications made by: You
- Owner's name: Your teacher; address and phone may be fictitious
- Builder's name: You, your address and phone number
- You sign as applicant
- Homebuilder's registration no.: Your social security number or another fictitious number
- Date: Today's date

PROBLEM 17-2

BUILDING PERMIT APPLICATION

Amount Due _____

Project Location (Address) _____

Nearest Cross Street _____

Subdivision Name _____ Lot _____ Block _____

Township _____ Range _____ Section _____ Tax Lot _____

Lot Size _____ (Sq. Ft.) Building Area _____ (Sq. Ft.) Basement Area _____ (Sq. Ft.) Garage Area _____ (Sq. Ft.)

Stories _____ Bedrooms _____ Water Source _____ Sewage Disposal _____

Estimated Cost of Labor and Material _____

Plans and Specifications made by _____ accompany this application.

Owner's Name _____ Builder's Name _____

Address _____ Address _____

City _____ State _____ City _____ State _____

Phone _____ Zip _____ Phone _____ Zip _____

I certify that I am registered under the provisions of ORS Chapter 701 and my registration is in full force and effect. I also agree to build according to the above description, accompanying plans and specifications, the State of Oregon Building Code, and to the conditions set forth below.

_____ _____ _____
APPLICANT HOMEBUILDER'S REGISTRATION NO. DATE

I agree to build according to the above description, accompanying plans and specifications, the State of Oregon Building Code, and to the conditions set forth below.

_____ _____
APPLICANT DATE

TO BE FILLED IN BY APPLICANT

PROBLEM 17-3

Complete the building contract on pages 237 and 238 based on this information:

- Today's date.
- Name yourself as the contractor and your teacher as the owner. (Give complete first name, middle initial, and last name where names are required.)
- ARTICLE I:

  Construction of approximate 2000-square-foot house to be located at 3456 Barrington Drive, your city and state. Also known as Lot 8, Block 2, Barrington Heights, your county and state.
- Drawings and specifications prepared by you.
- All said work to be done under the direction of you.
- ARTICLE II:

  Commence work within 10 days and substantially complete on or before: Give a date four months from the date of the contract.
- ARTICLE III:

  $88,500

  Payable at the following times:

  1. One-third upon completion of the foundation.

  2. One-third upon completion of drywall.

  3. One-third 30 days after posting the completion notice.
- ARTICLE X:

  Insurance $250,000; $500,000; and $50,000.
- ARTICLE XVIII:

  Give the owner's name, address, and phone number. (A fictitious address and phone number may be used if preferred.) Give your name and address as the contractor.
- ARTICLE XX:

  This contract is valid for a period of 60 days. If, for reasons out of the contractor's control, construction has not begun by the end of this 60-day period, contractor has the right to rebid and revise the contract.

PROBLEM 17-3

FORM No. 144—BUILDING CONTRACT (Fixed Price—No Service Charge).

TN

THIS AGREEMENT, Made the ....day of ...., 19........., by and between ...., hereinafter called the Contractor, and ...., hereinafter called the Owner, WITNESSETH:

    The parties hereto, each in consideration of the promises of the other, agree as follows:

    ARTICLE I: The contractor shall and will perform all the work for the

as shown on the drawings and described in the specifications therefor prepared by ....
....;
said drawings, specifications and this contract hereinafter, for brevity, are called "contract documents"; they are identified by the signatures of the parties hereto and hereby are made a part hereof. All said work is to be done under the direction of ....
....who, for brevity hereinafter is designated as "supervisor." (Publisher's note: If the owner himself is to supervise said work, simply insert the word "owner" in the blank space immediately preceding.) The supervisor's decision as to the true construction and meaning of the drawings and specifications shall be final and binding upon both parties. All of said drawings and specifications including those hereinafter mentioned have been and will be prepared by the owner at his expense and are to remain his property; said drawings and specifications are loaned to the contractor for the purposes of this contract and at the completion of the work are to be returned to the owner; none of said contract documents shall be used by, submitted or shown to third parties without owner's written consent.

    ARTICLE II: The contractor shall commence work within .... days from the date hereof and substantially complete the same on or before ...., 19 .... At all times the supervisor shall have access to said work for the purpose of inspecting the same and the progress thereof. Should completion be delayed by reason of the fault of the owner or of any other contractor employed by him or by fire, casualty, strikes, delays in obtaining materials or other reasons beyond the contractor's control, then the completion date shall be extended for a period equivalent to the time lost for such reasons. Should the parties be unable to agree as to the period of such extension, the question shall be referred to arbitration as hereunder provided. However, the contractor shall take special precautions to protect his work during freezing weather and shall be fully responsible for the effect of such weather upon said work.

    ARTICLE III: Subject to the provisions for adjustment set forth in ARTICLE V hereof, the owner shall pay to the contractor for the performance of this contract, in current funds, the sum of $ ...., payable at the following times:

NOTE—This form not suitable for use as a retail installment contract where a finance charge is being made.

Courtesy Stevens-Ness Law Publishing Co.

PROBLEM 17-3 (CONTINUED)

ARTICLE XII: The contractor shall keep the premises (especially that part thereof under the floors thereof) free from accumulation of waste materials or rubbish and at the completion of the work shall remove all of his tools, scaffoldings and supplies and leave the premises broom-clean, or its equivalent.

ARTICLE XIII: If the owner should require a completion bond from the contractor, the premium therefor shall be added to the contract price and paid by the owner on delivery of said bond to him.

ARTICLE XIV: If the contractor employs a foreman or superintendent on said work, all directions and instructions given to the latter shall be as binding as if given to the contractor.

ARTICLE XV: The contractor agrees at all times to keep said work and the real estate on which the same is to be constructed free and clear of all construction and materialmen's liens, including liens on behalf of any subcontractor or person claiming under any such subcontractor and to defend and save the owner harmless therefrom.

ARTICLE XVI: In all respects the contractor shall be deemed to be an independent contractor.

ARTICLE XVII: In the event of any suit or action arising out of this contract, the losing party therein agrees to pay to the prevailing party therein the latter's costs and reasonable attorney's fees to be fixed by the trial court and in the event of an appeal, the prevailing party's costs and reasonable attorney's fees in the appellate court to be fixed by the appellate court.

ARTICLE XVIII: Any notice given by one party hereto to the other shall be sufficient if in writing, contained in a sealed envelope with postage thereon fully prepaid and deposited in the U. S. Registered Mails; any such notice conclusively shall be deemed received by the addressee thereof on the day of such deposit. If such notice is intended for the owner, the envelope containing the same shall be addressed to the owner at the following address: ....

and if intended for the contractor, if addressed to ....
....

ARTICLE XIX: In construing this contract and where the context so requires, the singular shall be deemed to include the plural, the masculine shall include the feminine and the neuter and all grammatical changes shall be made and implied so that this contract shall apply equally to individuals and to corporations; further, the word "work" shall mean and include the entire job undertaken to be performed by the contractor as described in the contract documents, and each thereof, together with all services, labor and materials necessary to be used and furnished to complete the same, except for the preparation of the said plans and specifications and further except the compensation of the said supervisor.

ARTICLE XX: The parties hereto further agree

IN WITNESS WHEREOF, the parties have hereunto set their hands in duplicate.

....
CONTRACTOR

....

....
OWNER

....

PROBLEM 18-4

Draw cedar shakes on the roof, with 1" exterior stucco on all walls. Show shadows by placing the sun in the upper-left-hand corner.

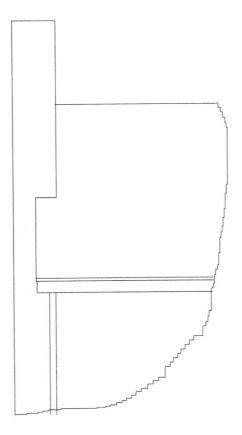

PROBLEM 18-5

Complete the drawing below by adding 4'-'0" x 3'-6" sliding window. Show the roof with clay tiles. Use wood shingles above and below the windows, with 1 x 4 trim. Other siding to be 1" exterior stucco. Determine the location of the sun and show all shadows.

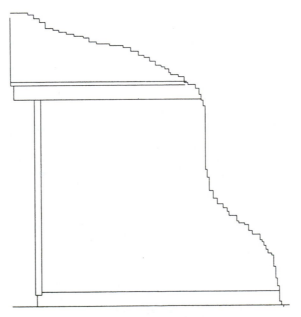

PROBLEM 18-6

Show the shadows for a 24" cantilever and a 24" eave overhang. Show horizontal siding and composition roofing.

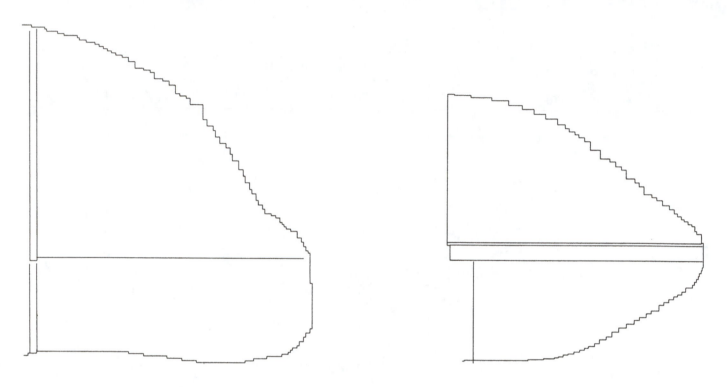

PROBLEM 18-7

Complete the chimneys below using wood shingles on the left and brick on the right. Determine a light source and show all shadows.

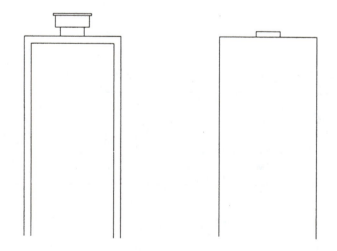

PROBLEM 18-8

Show the shadows that would result from a 24" and a 30" overhang.

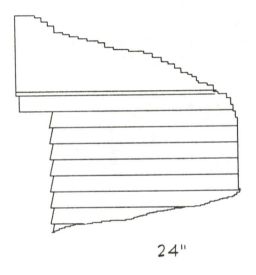

24"

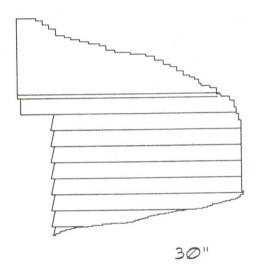

30"

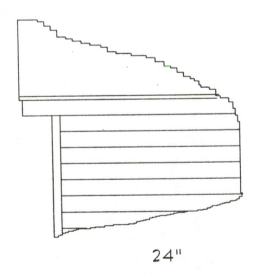

24"

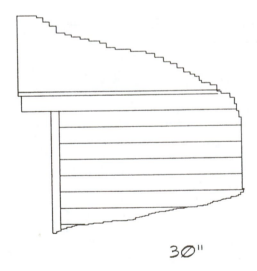

30"

PROBLEM 18-9

Show the shadows that would result from 24" overhangs and 12" overhangs at the gable ends. Use a scale of 1/8" = 1'-0".

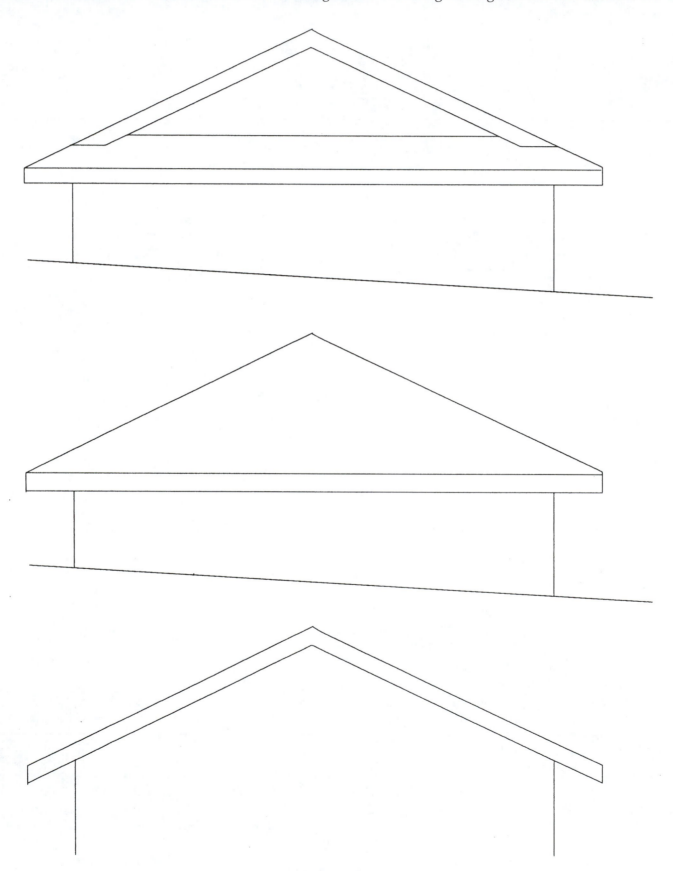

PROBLEM 18-10

Use the floor plan below and show a driveway, stone walkway, water fountain, and garden in the front yard. Show a minimum of two different kinds of trees and two different types of shrubs. Use a scale of 3/16" = 1'-0".

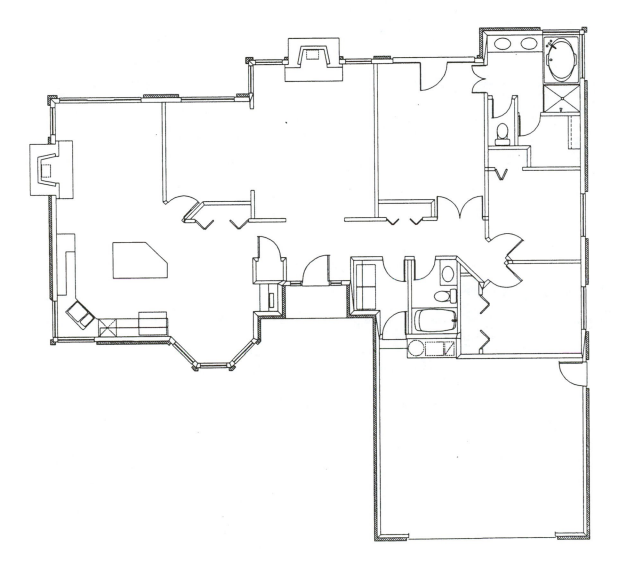

PROBLEM 18-11

Complete the presentation section shown below by including two types of trees and shrubs. Show a winter sun angle of 22 degrees and a summer sun angle of 68 degrees.

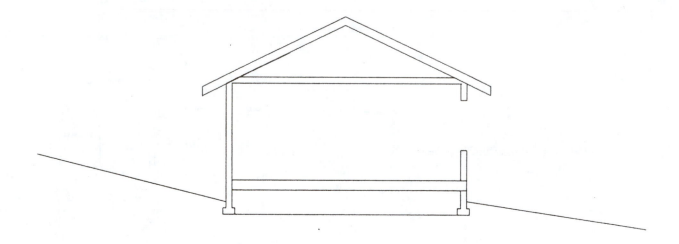

PROBLEM 18-12

Complete the presentation section shown below by including two types of trees and shrubs. Show a winter sun angle of 24 degrees and a summer sun angle of 66 degrees.

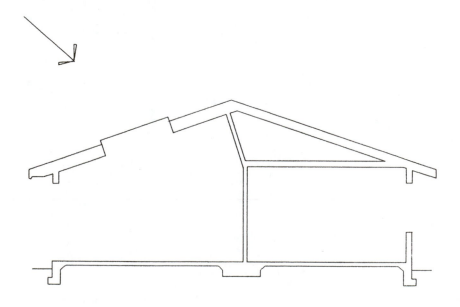

PROBLEM 18-13

Using your drawing from Chapter 2 as a guide, complete a presentation site plan. Show a minimum of three different trees, walks, a pool, and a garden area.

PROBLEM 18-14

Using the floor plan from Chapter 4, render a floor plan. Show a minimum of three styles of plantings, decks, and furniture.

PROBLEM 18-15

Using the elevation that was drawn in Chapter 9 as a base, draw a presentation drawing.

PROBLEM 18-16

Using the section that was drawn in Chapter 13 as a guide, draw a presentation drawing. Show trees, shrubs, and sun angles appropriate for your area.

PROBLEM 18-17

Mount a copy of your floor plan and elevations on illustration board.

PROBLEM 18-18

Draw a two-point perspective, placing the interior corner on the PP so that part of the residence will extend below the PP.

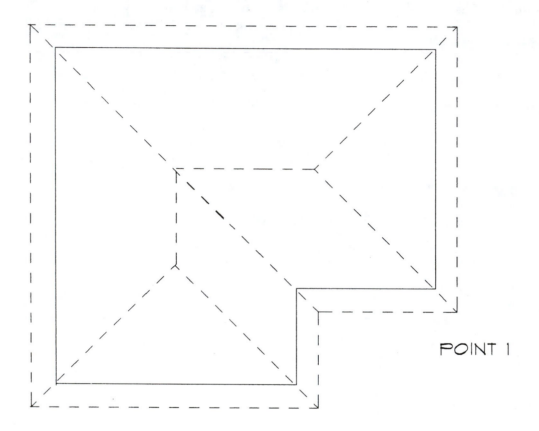

POINT 1

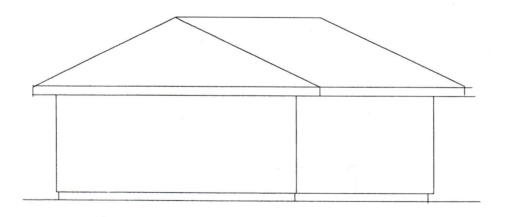

PROBLEM 18-19

Draw a perspective drawing that will give the best view of the courtyard.

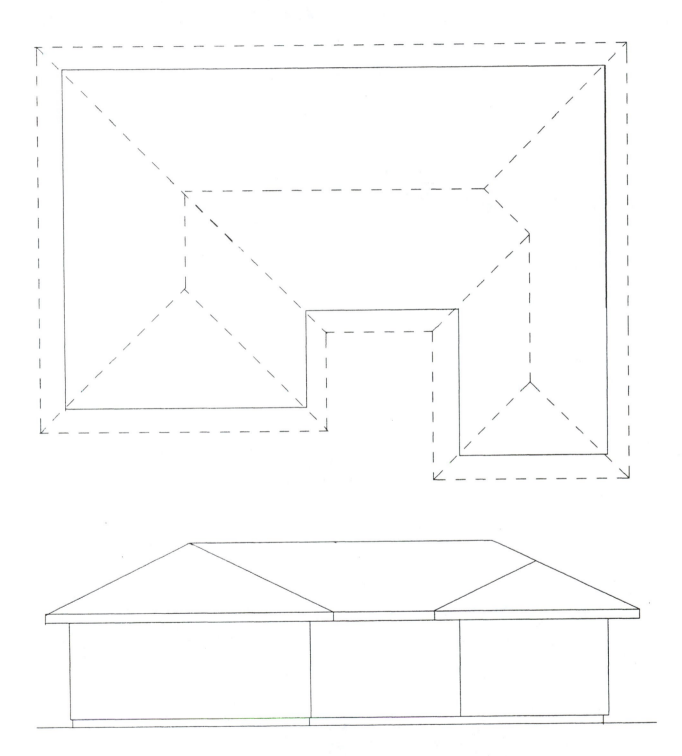

PROBLEM 18-20

Draw a two-point perspective that will accent the short side of the residence. Use standard heights for all materials on the floor plans. Use a scale of 1/2" = 1'-0" unless noted. Assume an eight foot ceiling.

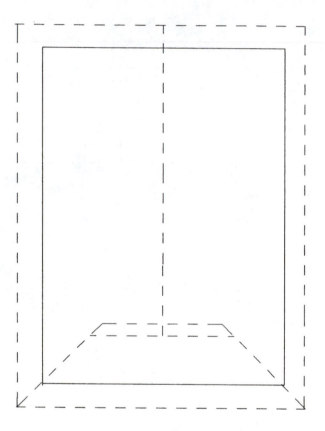

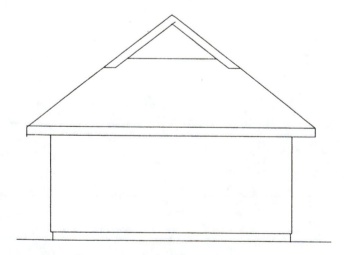

PROBLEM 18-21

Draw a perspective drawing from points A and B. Use the same horizon lines.

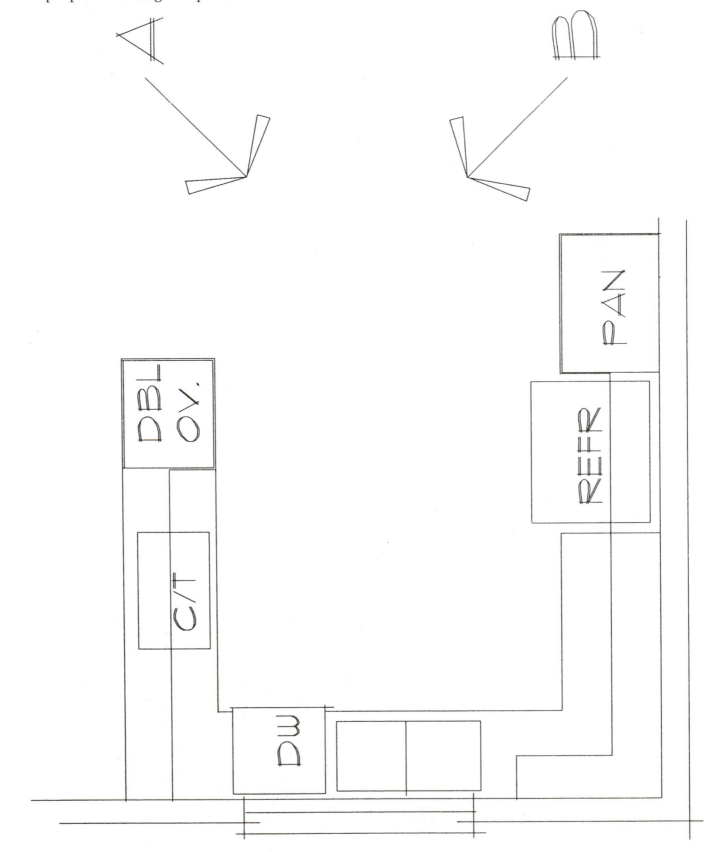

SECTION

Advanced Residential Projects

DIRECTIONS These problems allow you to design floor plans based on preliminary layouts. You may select or your instructor will assign one or more floor plan problems depending on your course objectives. Each problem is presented as a preliminary design layout. Using the preliminary design as a guide, do the following:

- Design and draw the floor plan to fit one of the sites in Chapter 2.
- Design all window, skylight, and door sizes and types based on or exceeding minimum code requirements.
- Display the window size and type and the door sizes on the floor plan or in a schedule, depending on your course objectives.
- Completely label and provide all related floor plan general and local notes.
- Show and label all construction members on a separate framing plan.
- When stairs are used, calculate and accurately show the number and size of risers.
- Design and show all appliances and plumbing fixtures.
- Label all wardrobes, closets, pantries, and related storage.
- Use appropriate titles, scales, and title blocks.
- Design the complete electrical system and draw with the floor plan or as a separate Electrical Plan.
- Design and draw the roof plan.
- Design and draw all elevations.
- Design and draw the complete foundation plan.
- Draw all related sections and details.
- Most structural dimensions are given on the floor plan or on framing plan reference sheets for more complex homes. Calculate and label all missing structural members.

This series of problems leads you to the completion of a set of architectural working drawings for one or more homes. It is recommended that you pick a specific plan and follow it through the entire set of working drawings as you continue into advanced chapters of architectural Drafting & Design.

PROBLEM 1

Using the floor plans on the following pages and the elevation below, design and draw the required elevations using saltbox, garrison, or Victorian styling. Use a scale of 1/4" = 1'-0" for each elevation unless your instructor gives other instructions.

FRONT ELEVATION

SCALE : 1/4" = 1'-0"

PROBLEM 1 (CONTINUED)

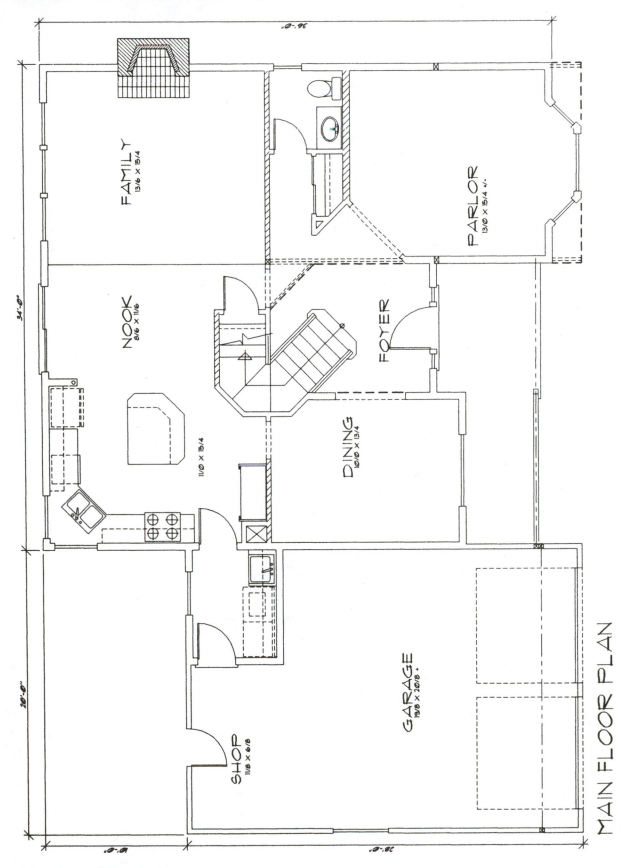

Courtesy of Alan Mascord Design Associates.

PROBLEM 1 (CONTINUED)

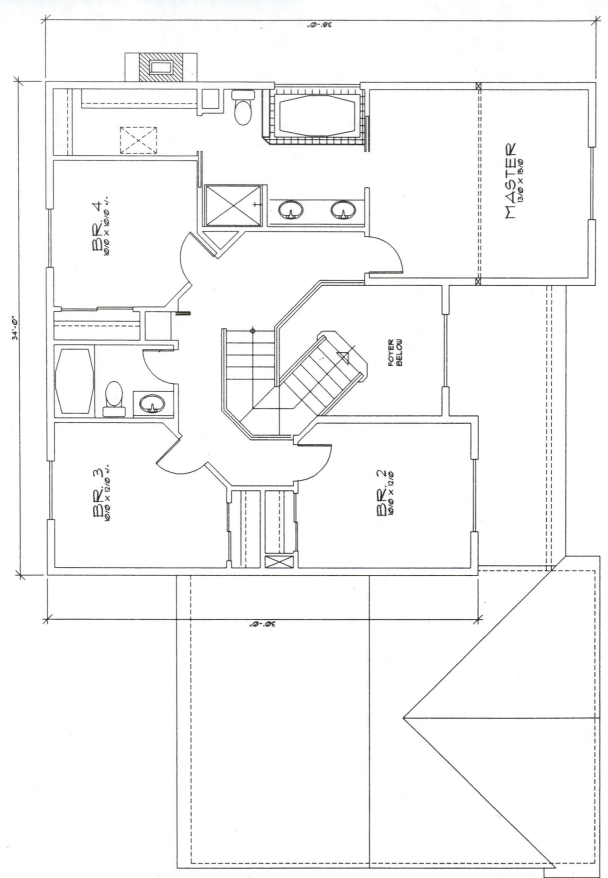

SECOND FLOOR PLAN

PROBLEM 1 (CONTINUED)

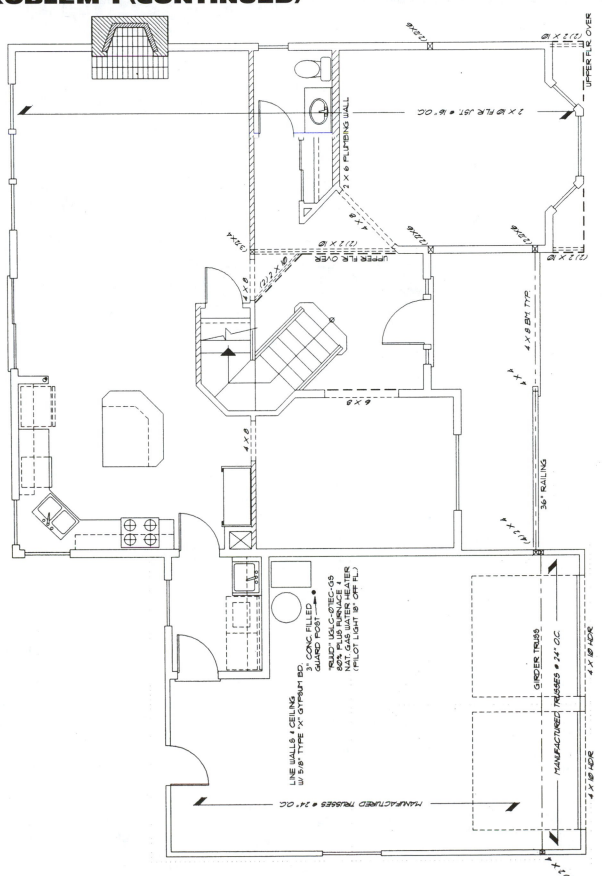

MAIN FLOOR
FRAMING PLAN REFERENCE

PROBLEM 1 (CONTINUED)

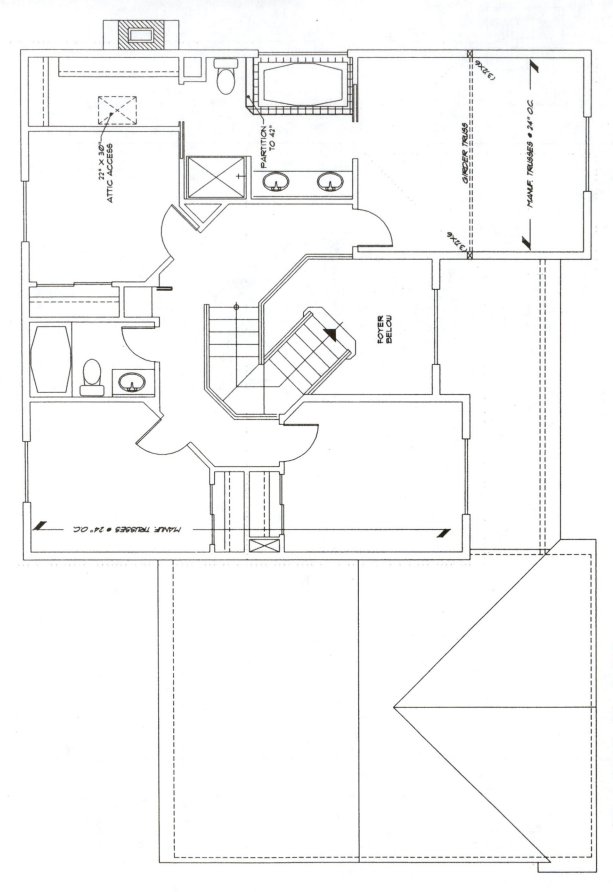

SECOND FLOOR
FRAMING PLAN REFERENCE

PROBLEM 2

Using the floor plans on the following pages and the elevation below, design and draw the required elevations using Spanish, Georgian, or Victorian styling.

FRONT ELEVATION
SCALE: 1/4" x 1' - 0"

PROBLEM 2 (CONTINUED)

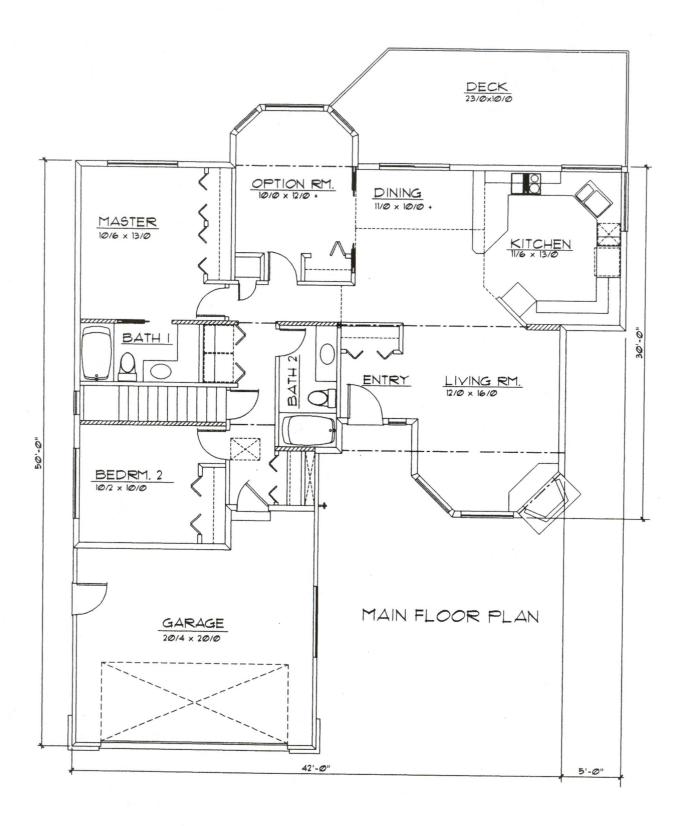

DECK
23/0 x 10/0

OPTION RM.
10/0 x 12/0 +

DINING
11/0 x 10/0 +

MASTER
10/6 x 13/0

KITCHEN
11/6 x 13/0

BATH 1

BATH 2

ENTRY

LIVING RM.
12/0 x 16/0

BEDRM. 2
10/2 x 10/0

GARAGE
20/4 x 20/0

MAIN FLOOR PLAN

50'-0"

30'-0"

42'-0"

5'-0"

Courtesy of Piercy & Barclay Designers, Inc.

PROBLEM 2 (CONTINUED)

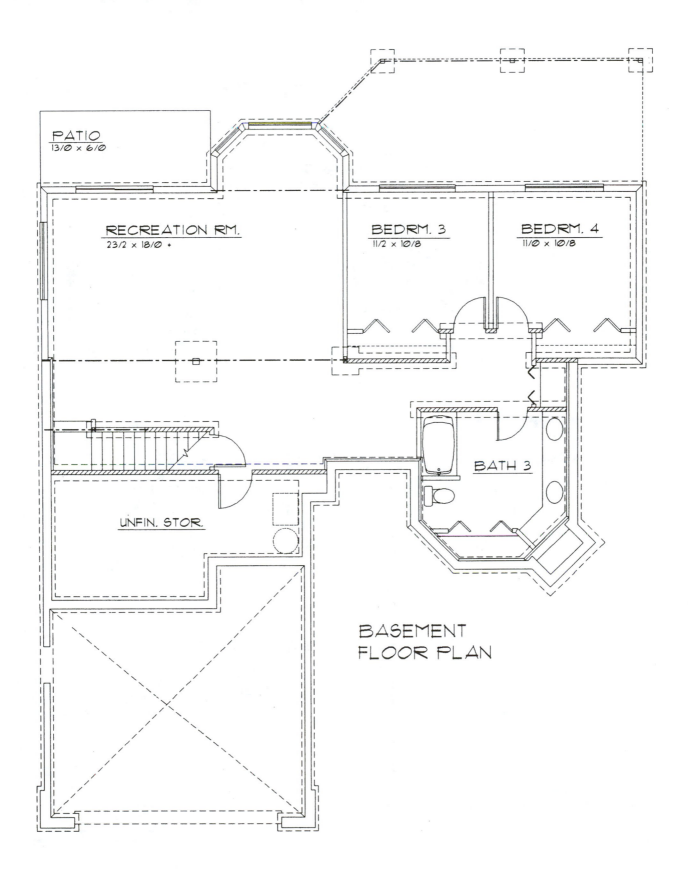

PATIO
13/0 × 6/0

RECREATION RM.
23/2 × 18/0 +

BEDRM. 3
11/2 × 10/8

BEDRM. 4
11/0 × 10/8

BATH 3

UNFIN. STOR.

BASEMENT
FLOOR PLAN

PROBLEM 2 (CONTINUED)

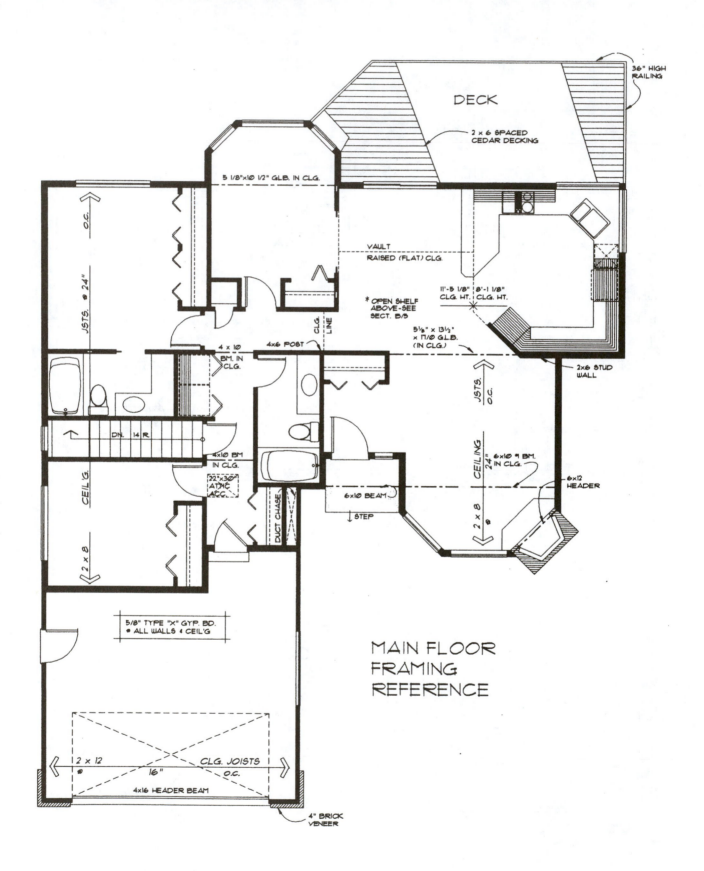

DECK

36" HIGH
RAILING

2 x 6 SPACED
CEDAR DECKING

5 1/8"x10 1/2" GLB. IN CLG.

VAULT
RAISED (FLAT) CLG.

11'-5 1/8"   8'-1 1/8"
CLG. HT.   CLG. HT.

* OPEN SHELF
ABOVE-SEE
SECT. B/3

5⅛" x 13½"
x 17/0 GLB.
(IN CLG.)

JSTS. @ 24"

CLG.
LINE

O.C.

JSTS. O.C.

2x6 STUD
WALL

4 x 10
BM. IN
CLG.

4x6 POST

JSTS. @ 24"

CEILING 24"

6x10 "9 BM.
IN CLG.

6x12
HEADER

4x10 BM
IN CLG.

22"x30"
ATTIC
ACC.

6x10 BEAM

↓ STEP

2 x 8

CEIL'G.

DUCT CHASE

2 x 8

DN. 14 R

5/8" TYPE "X" GYP. BD.
● ALL WALLS & CEIL'G

MAIN FLOOR
FRAMING
REFERENCE

2 x 12   CLG. JOISTS
16"   O.C.

4x16 HEADER BEAM

4" BRICK
VENEER

PROBLEM 2 (CONTINUED)

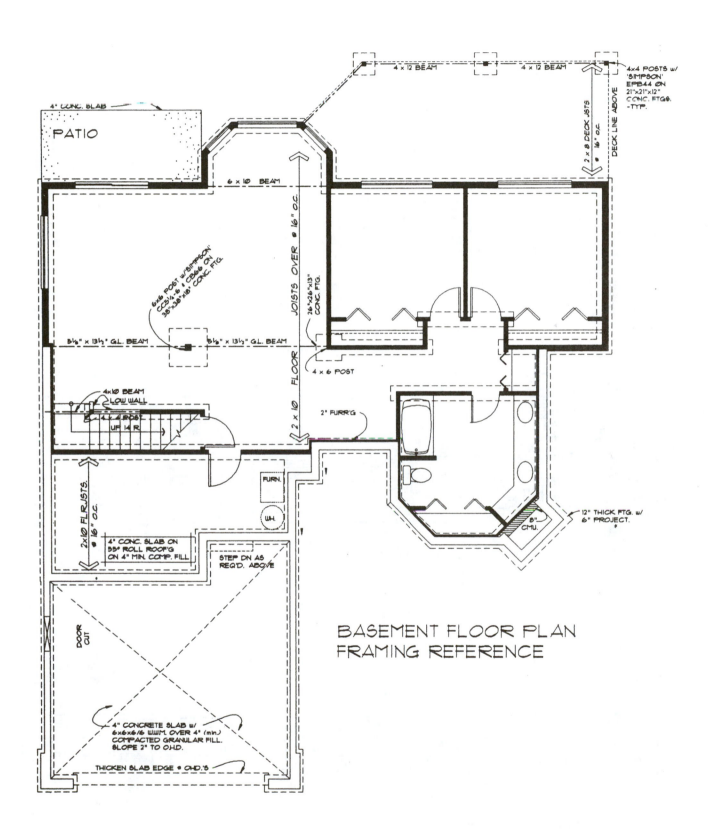

BASEMENT FLOOR PLAN
FRAMING REFERENCE

PROBLEM 3

Using the floor plans on the following pages and the elevation below, design and draw the required elevations using French or Dutch colonial or Victorian styling.

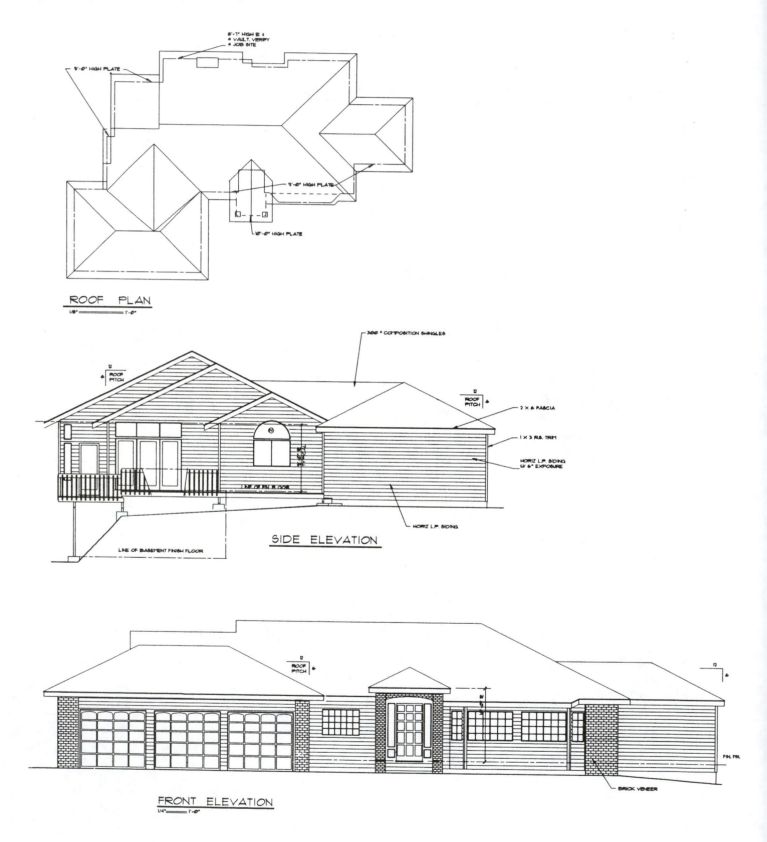

ROOF PLAN

SIDE ELEVATION

FRONT ELEVATION

PROBLEM 3 (CONTINUED)

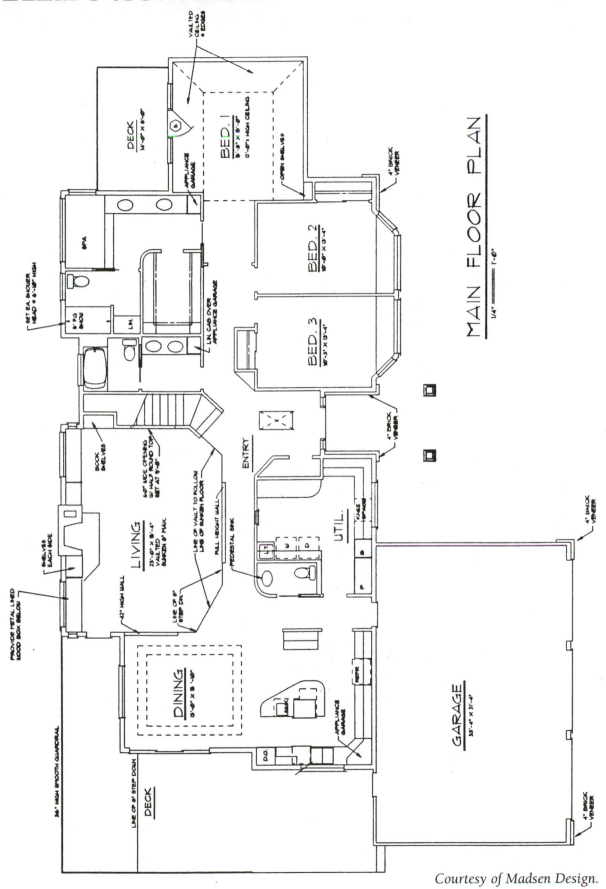

Courtesy of Madsen Design.

PROBLEM 3 (CONTINUED)

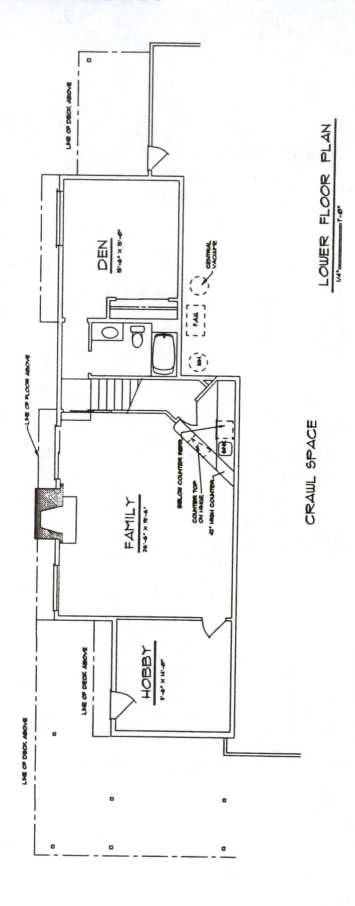

LOWER FLOOR PLAN
1/4" = 1'-0"

PROBLEM 4

Using the floor plans on the following pages and the elevations below, design and draw the required elevations. Use French colonial, English tudor, Georgian, half timber, or Spanish styling.

FRONT ELEVATION

PROBLEM 4 (CONTINUED)

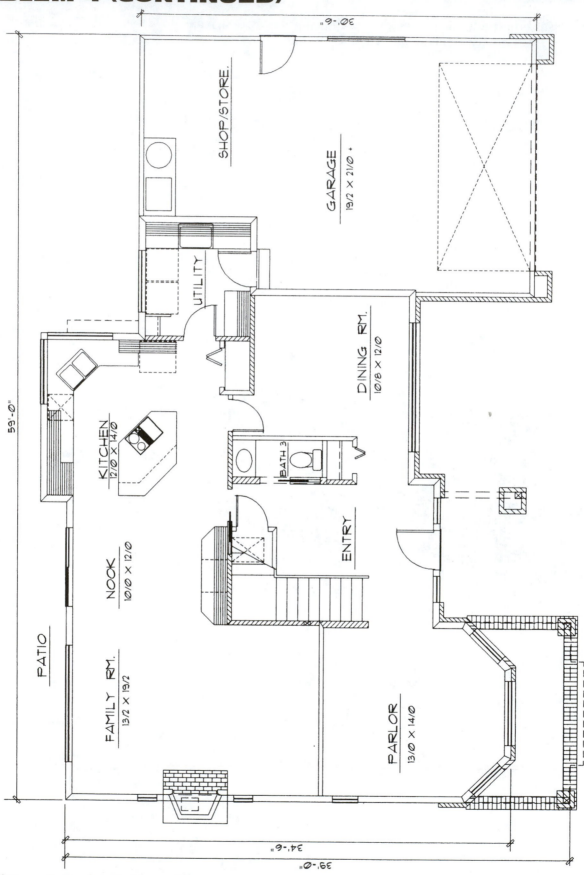

Courtesy of Piercy & Barclay Designers, Inc.

PROBLEM 4 (CONTINUED)

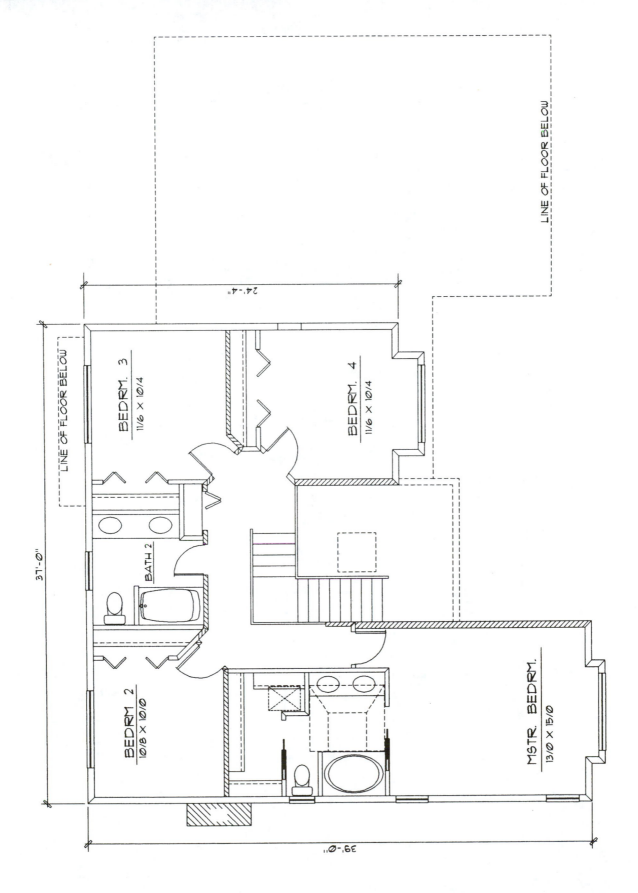

PROBLEM 4 (CONTINUED)

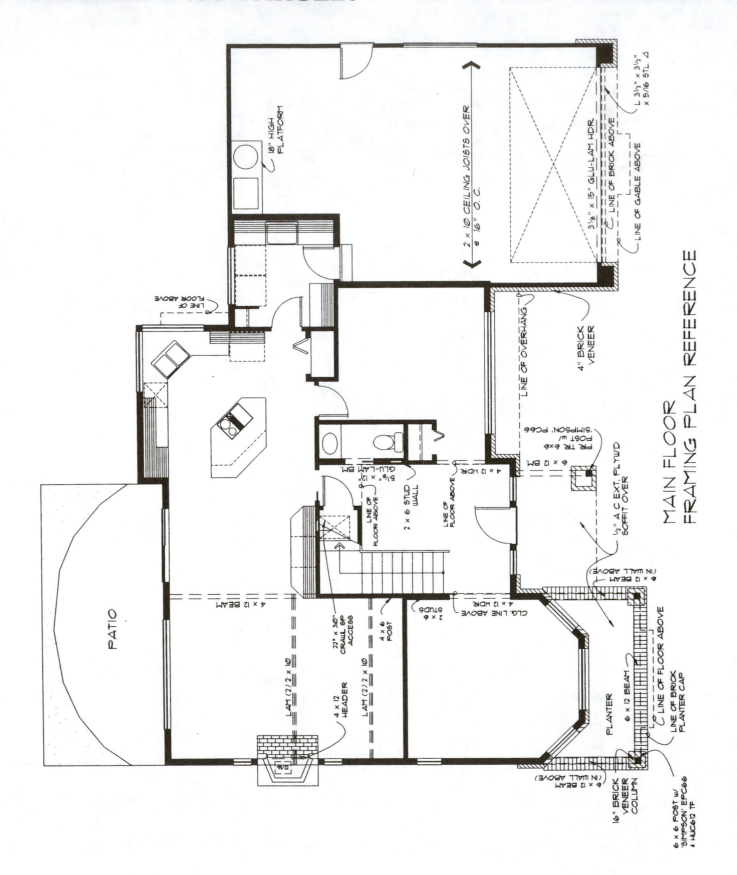

MAIN FLOOR
FRAMING PLAN REFERENCE

PROBLEM 4 (CONTINUED)

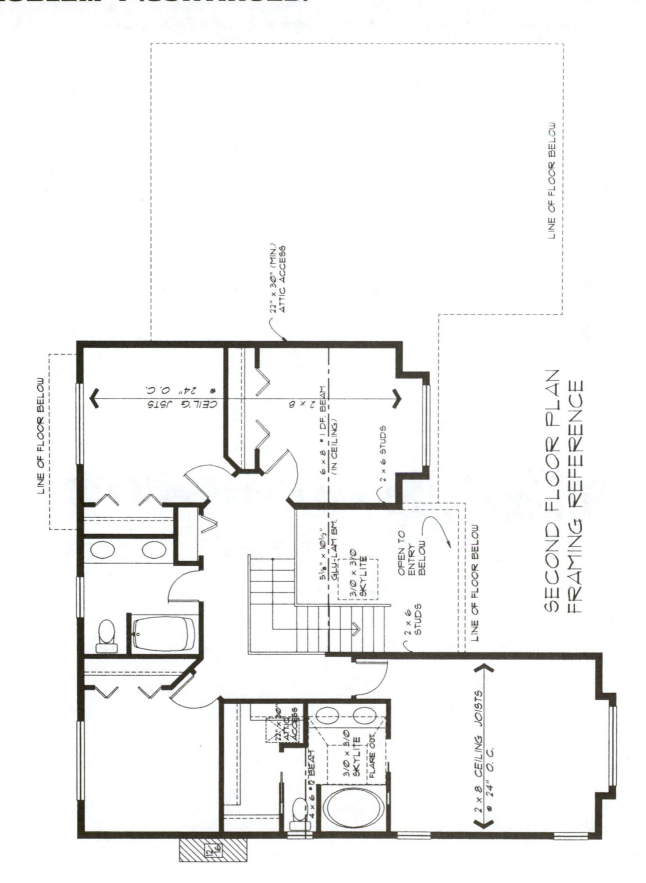

SECOND FLOOR PLAN
FRAMING REFERENCE

PROBLEM 5

Using the floor plans on the following pages and the elevation below, design and draw the required elevations. Pick siding and roofing appropriate for your area. Draw the front elevation as a presentation elevation and the other elevations as working.

FRONT ELEVATION

PROBLEM 5 (CONTINUED)

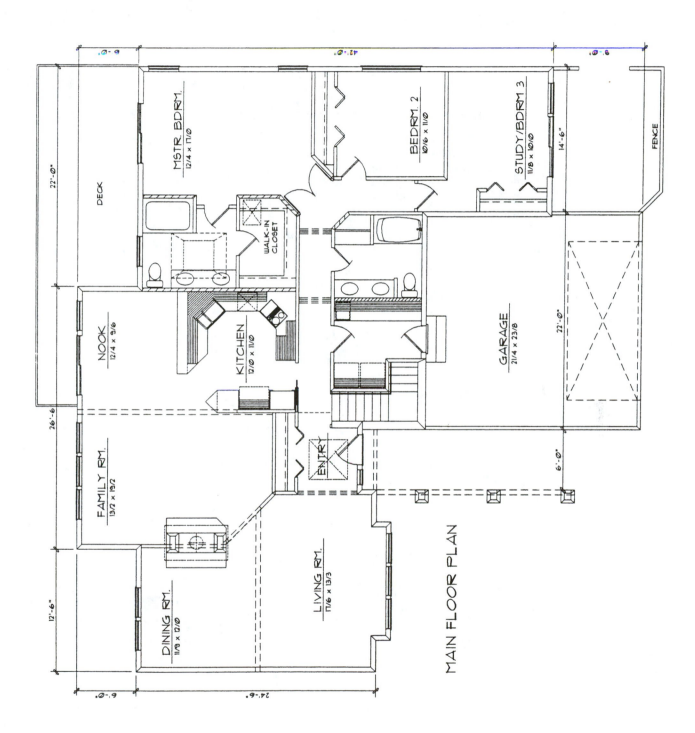

MAIN FLOOR PLAN

Courtesy of Piercy & Barclay Designers, Inc.

PROBLEM 5 (CONTINUED)

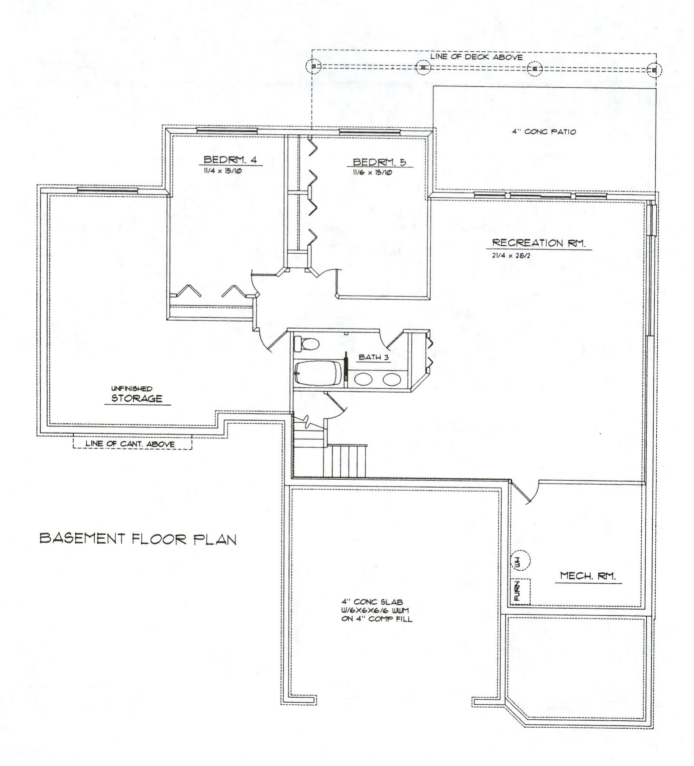

LINE OF DECK ABOVE

4" CONC PATIO

BEDRM. 4
11/4 x 15/10

BEDRM. 5
11/6 x 15/10

RECREATION RM.
21/4 x 28/2

BATH 3

UNFINISHED
STORAGE

LINE OF CANT. ABOVE

BASEMENT FLOOR PLAN

MECH. RM.

FURN

H W

4" CONC SLAB
W/6X6X6/6 WWM
ON 4" COMP FILL

PROBLEM 5 (CONTINUED)

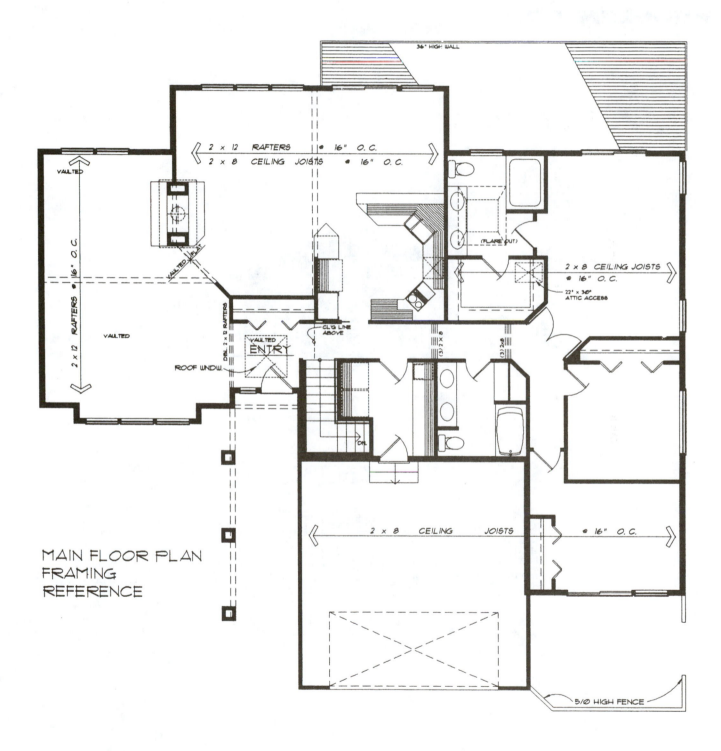

MAIN FLOOR PLAN
FRAMING
REFERENCE

DIRECTIONS: For Problems 6–10, use the floor plans and elevations given for each problem to draw the required elevations. Pick siding and roofing appropriate for your area.

PROBLEM 6(A)

FRONT ELEVATION

LEFT-SIDE ELEVATION

PROBLEM 6(A) (CONTINUED)

REAR ELEVATION

RIGHT-SIDE ELEVATION

PROBLEM 6(A) (CONTINUED)

Problem 6(a) is a dual master bedroom plan. A design of this type might be used as a shared home, with separate bedroom suites and joint living facilities.

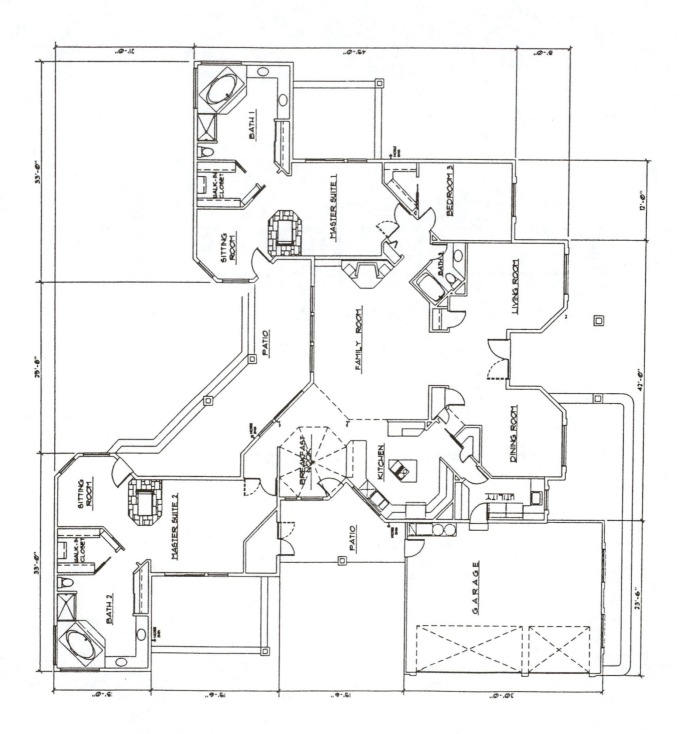

Courtesy of Piercy & Barclay Designers, Inc.

PROBLEM 6(B)

Using the floor plan in Problem 6(a) as a guide, redesign one of the master bedroom suites into a three-bedroom wing. This redesign will result in a four-bedroom home with a master bedroom suite and a children's bedroom wing. There should be at least one bathroom accessible to the three bedrooms in the children's wing.

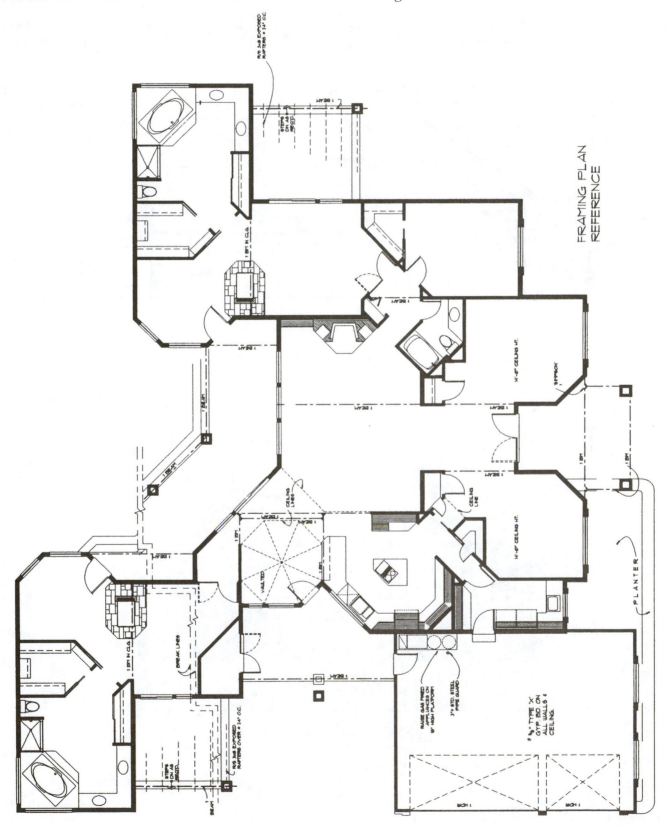

PROBLEM 7

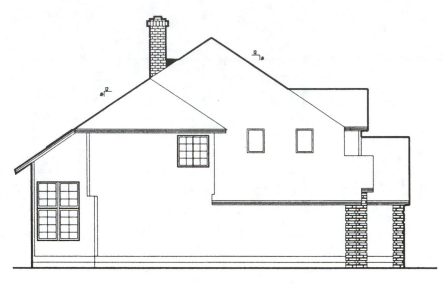

LEFT ELEVATION

FRONT ELEVATION

PROBLEM 7 (CONTINUED)

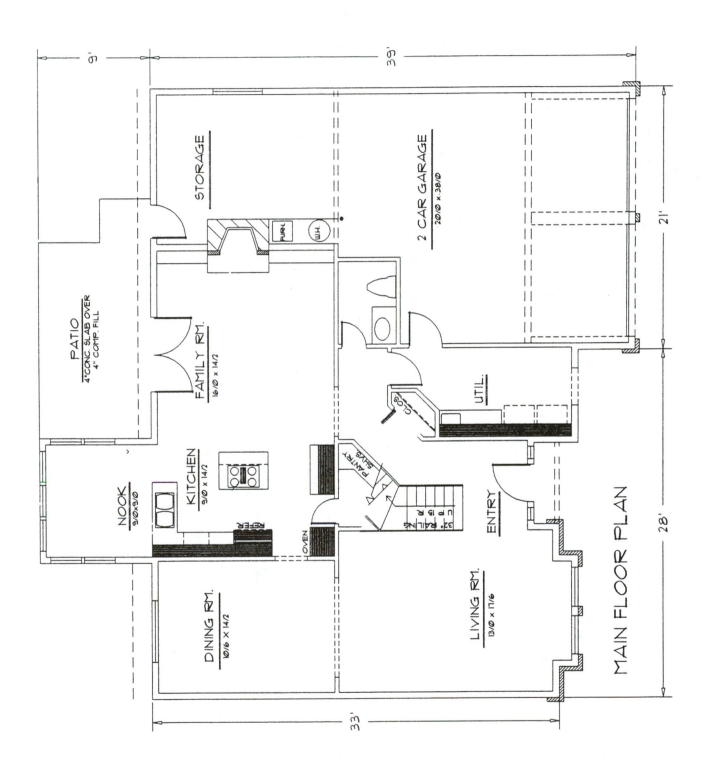

Courtesy of Sunridge Design, Wally Greiner AIBD.

PROBLEM 7 (CONTINUED)

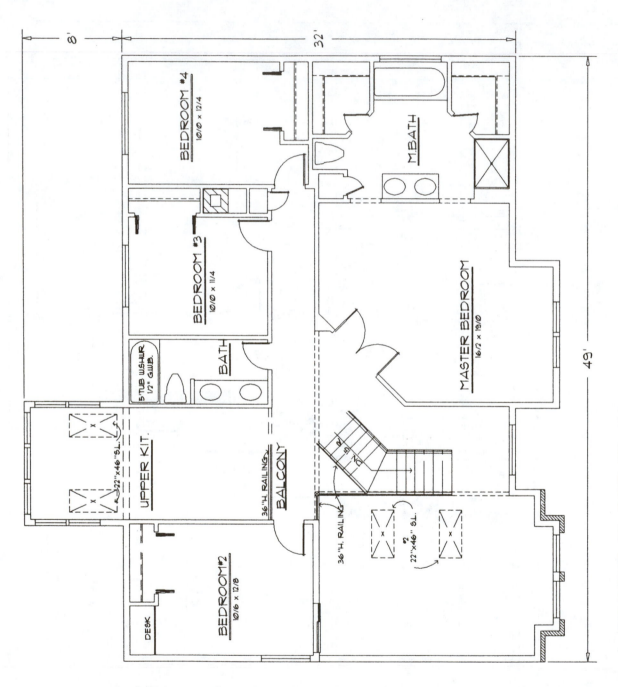

SECOND FLOOR PLAN

PROBLEM 7 (CONTINUED)

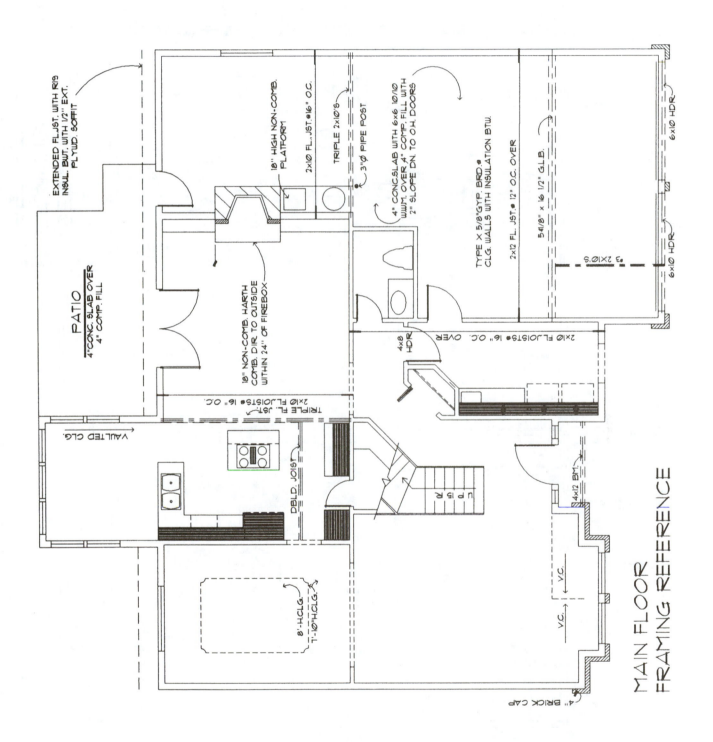

MAIN FLOOR
FRAMING REFERENCE

PROBLEM 7 (CONTINUED)

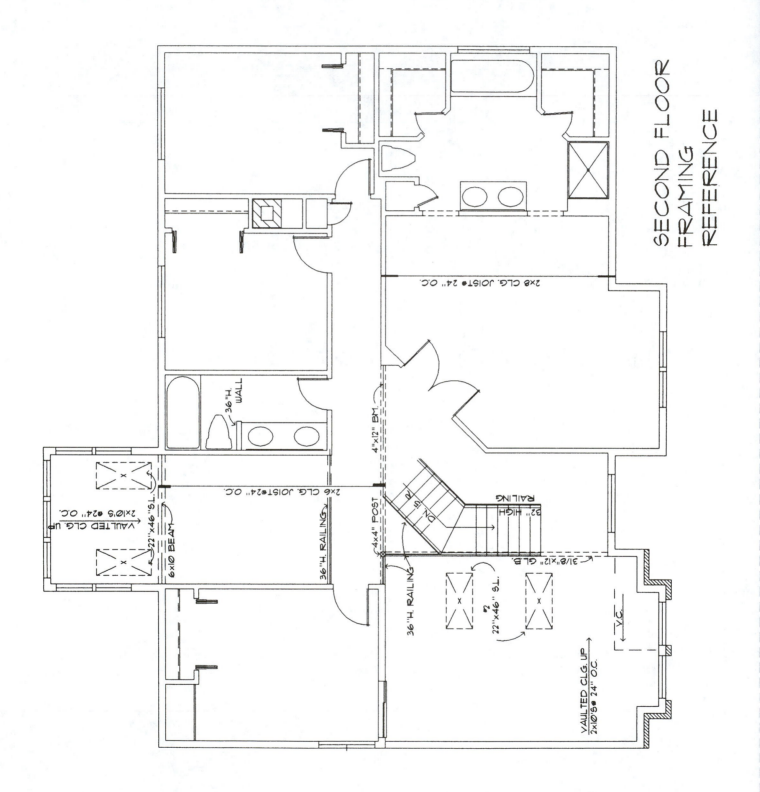

SECOND FLOOR FRAMING REFERENCE

PROBLEM 8

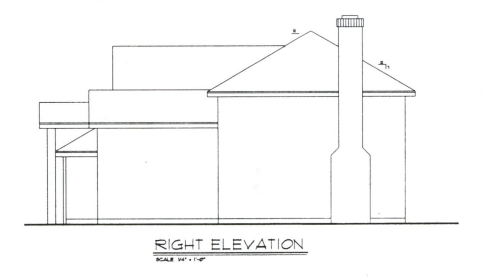

RIGHT ELEVATION
SCALE 1/4" = 1'-0"

FRONT ELEVATION
SCALE 1/4" = 1'-0"

PROBLEM 8 (CONTINUED)

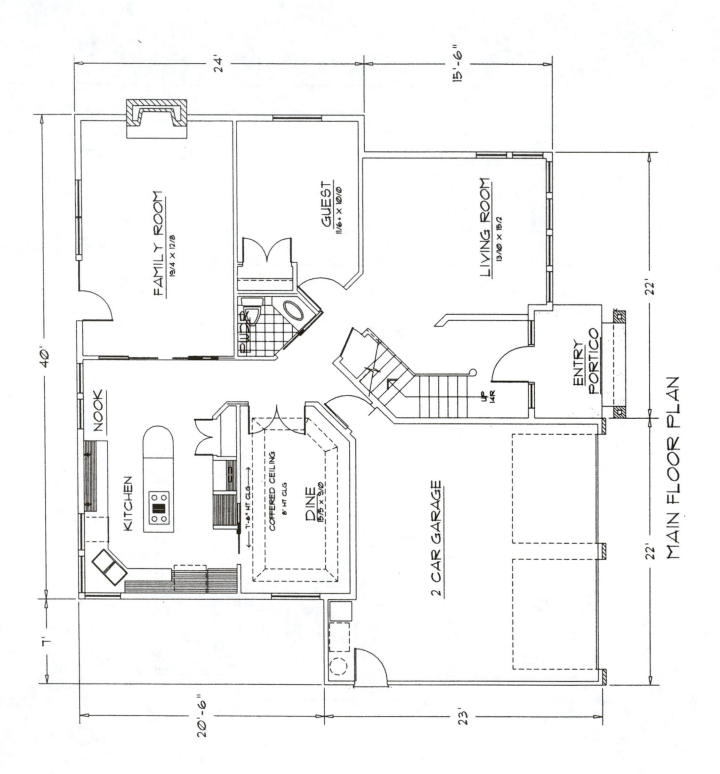

Courtesy of Sunridge Design, Wally Greiner AIBD.

PROBLEM 8 (CONTINUED)

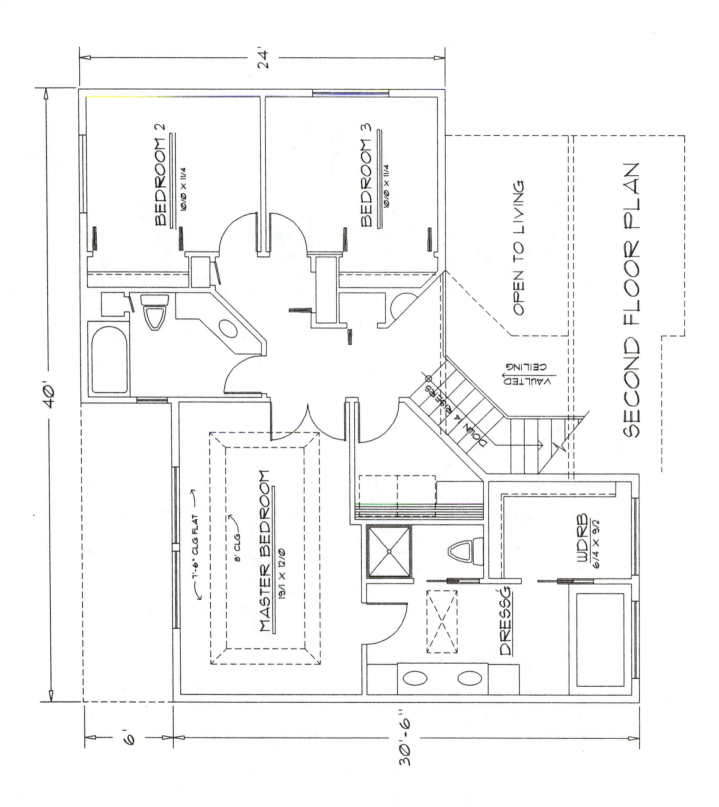

PROBLEM 8 (CONTINUED)

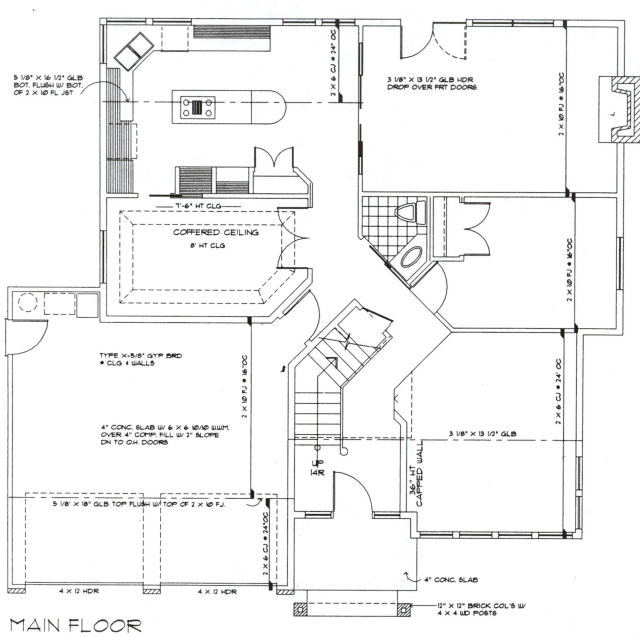

5 1/8" × 16 1/2" GLB
BOT. FLUSH W/ BOT.
OF 2 × 10 FL JST

2 × 6 CJ @ 24" OC

3 1/8" × 13 1/2" GLB HDR
DROP OVER FRT DOORS

2 × 10 FJ @ 16"OC

7'-6" HT CLG

COFFERED CEILING
8' HT CLG

2 × 10 FJ @ 16"OC

TYPE X-5/8" GYP BRD
@ CLG & WALLS

2 × 10 FJ @ 16"OC

UP
14R

2 × 6 CJ @ 24" OC

3 1/8" × 13 1/2" GLB

4" CONC. SLAB W/ 6 × 6 10/10 W.W.M.
OVER 4" COMP. FILL W/ 2" SLOPE
DN TO O.H. DOORS

36" HT
CAPPED WALL

5 1/8" × 18" GLB TOP FLUSH W/ TOP OF 2 × 10 FJ.

2 × 6 CJ @ 24"OC

4" CONC. SLAB

4 × 12 HDR   4 × 12 HDR

12" × 12" BRICK COL'S W/
4 × 4 WD POSTS

MAIN FLOOR
FRAMING REFERENCE

PROBLEM 8 (CONTINUED)

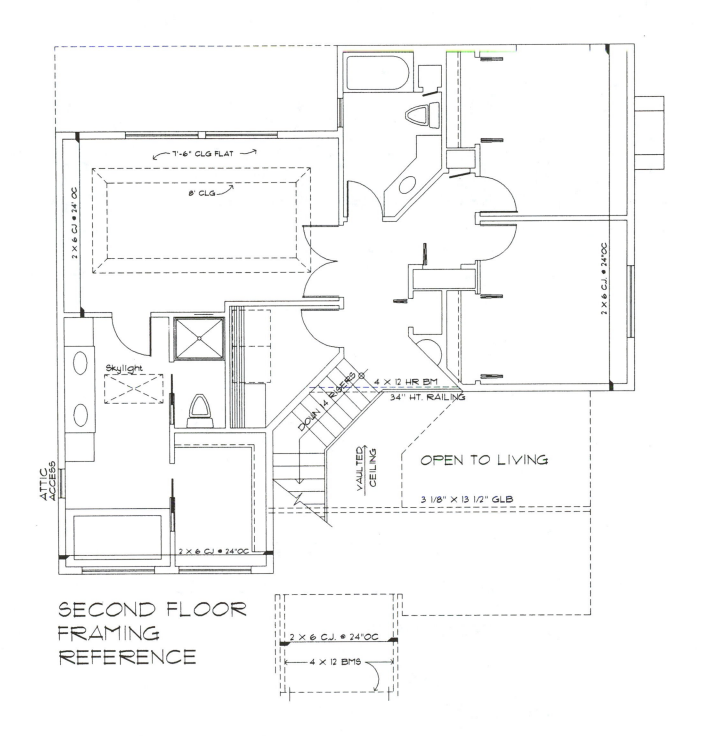

7'-6" CLG FLAT

8' CLG

2 x 6 CJ @ 24' OC

2 x 6 C.J. @ 24"OC

Skylight

ATTIC ACCESS

4 x 12 HR BM

DOWN 14 RISERS @

34" HT. RAILING

VAULTED CEILING

OPEN TO LIVING

3 1/8" X 13 1/2" GLB

2 x 6 CJ @ 24"OC

SECOND FLOOR
FRAMING
REFERENCE

2 x 6 C.J. @ 24"OC

4 x 12 BMS

PROBLEM 9

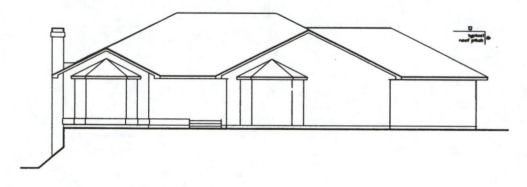

FRONT ELEVATION

RIGHT SIDE ELEVATION

PROBLEM 9 (CONTINUED)

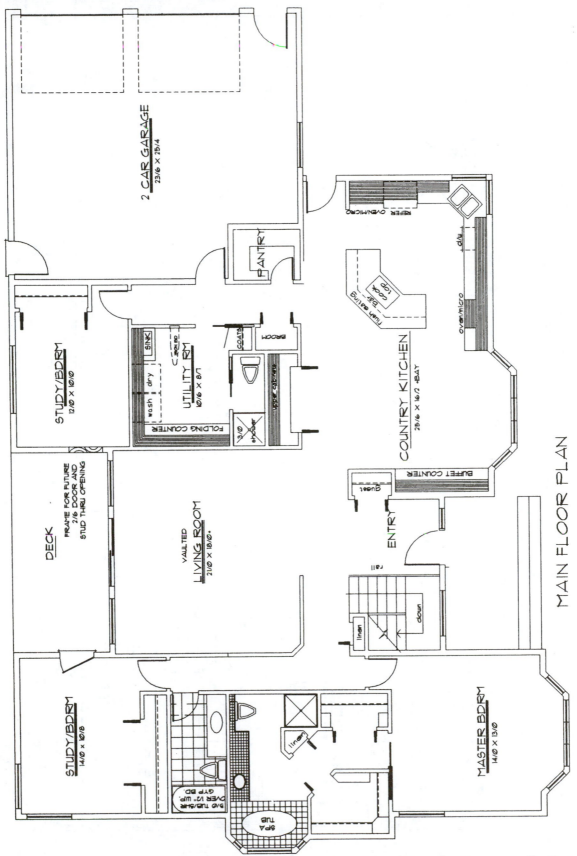

Courtesy of Sunridge Design, Wally Greiner AIBD.

PROBLEM 9 (CONTINUED)

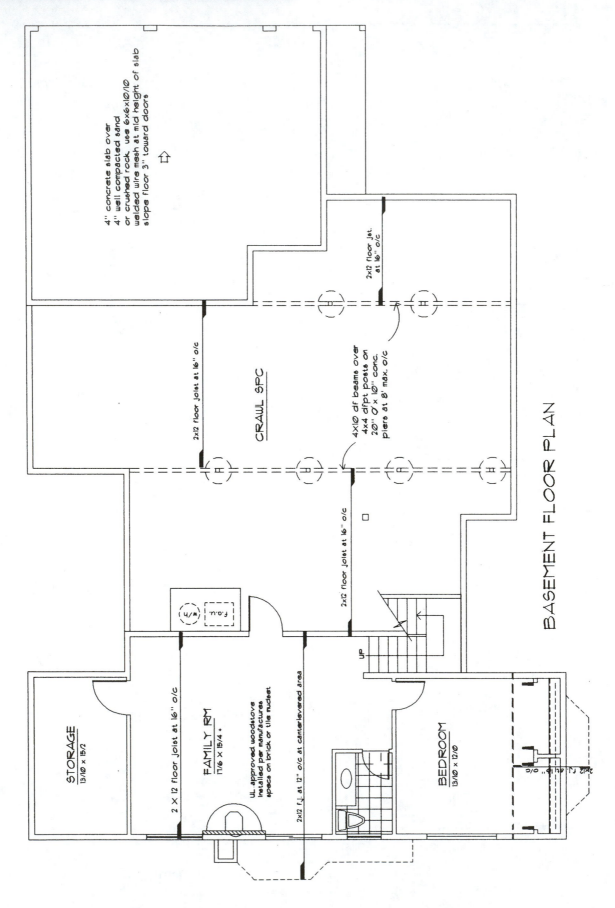

4" concrete slab over
4" well compacted sand
or crushed rock. use 6×6×10/10
welded wire mesh at mid height of slab
slopes floor 3" toward doors

2×12 floor jst.
at 16" o/c

2×12 floor joist at 16" o/c

CRAWL SPC

4×10 df beams over
4×4 dfpt posts on
20" Ø × 10" conc.
piers at 8' max. o/c

2×12 floor joist at 16" o/c

w/h   f.a.u.

UP

STORAGE
13/10 × 15/2

FAMILY RM
17/6 × 15/4 +

2 × 12 floor joist at 16" o/c

UL approved woodstove
installed per manufactures
specs on brick or tile nudset.

2×12 f.j. at 12" o/c at canterlevered area

BEDROOM
13/10 × 12/0

2×12 f.j. at 16" o/c

BASEMENT FLOOR PLAN

PROBLEM 9 (CONTINUED)

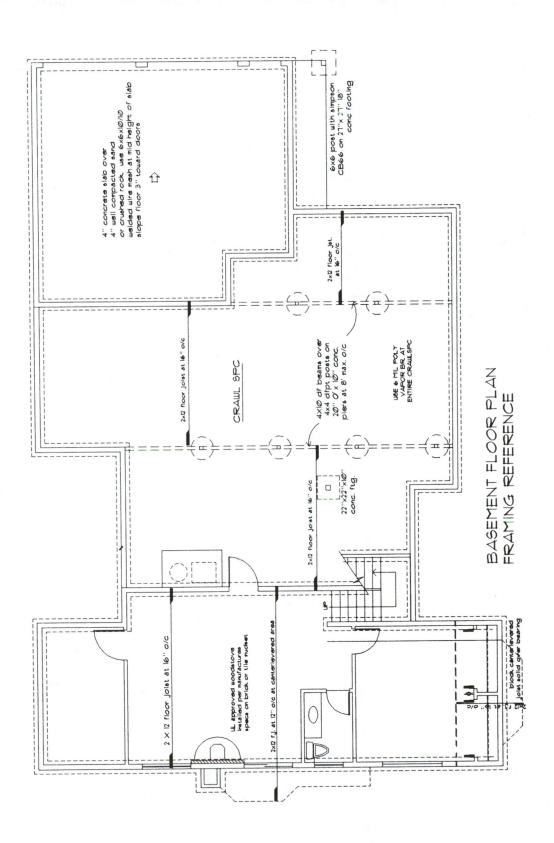

BASEMENT FLOOR PLAN
FRAMING REFERENCE

PROBLEM 10

Using the floor plans on the following pages and the elevation below, design and draw the required elevations. Pick siding and roofing appropriate for your area.

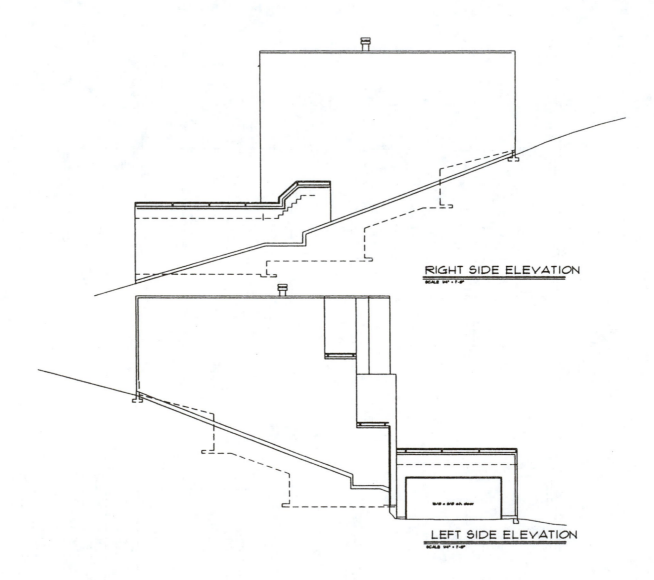

RIGHT SIDE ELEVATION
SCALE 1/4" = 1'-0"

LEFT SIDE ELEVATION
SCALE 1/4" = 1'-0"

PROBLEM 10 (CONTINUED)

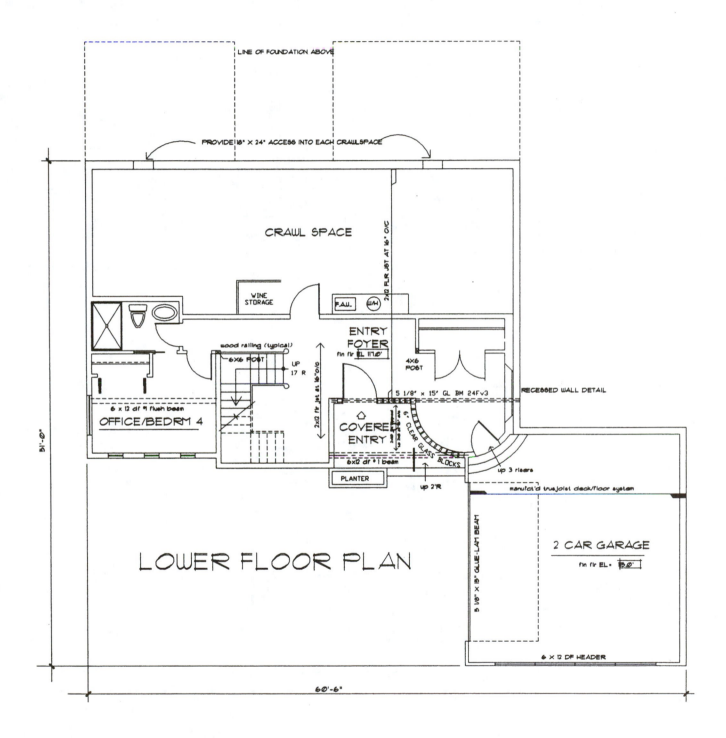

LINE OF FOUNDATION ABOVE

PROVIDE 18" X 24" ACCESS INTO EACH CRAWLSPACE

CRAWL SPACE

2x10 FLR JST AT 16" O/C

WINE STORAGE

F.A.U.   W/H

ENTRY FOYER
fin flr EL 117.0'

wood railing (typical)

6X6 POST

UP
17 R

4X6 POST

RECESSED WALL DETAIL

2x12 flr jst at 16"o/c

5 1/8" x 15' GL BM 24Fv3

6 x 12 df fl flush beam

OFFICE/BEDRM 4

COVERED ENTRY

6' CLEAR GLASS BLOCKS

up 3 risers

6x12 df fl beam

PLANTER

up 2'R

manufct'd trusjoist deck/floor system

5 1/8" x 15" GLUE-LAM BEAM

2 CAR GARAGE
fin flr EL = 130'

LOWER FLOOR PLAN

51'-0"

60'-6"

6 X 12 DF HEADER

Courtesy of Sunridge Design, Wally Greiner AIBD.

PROBLEM 10 (CONTINUED)

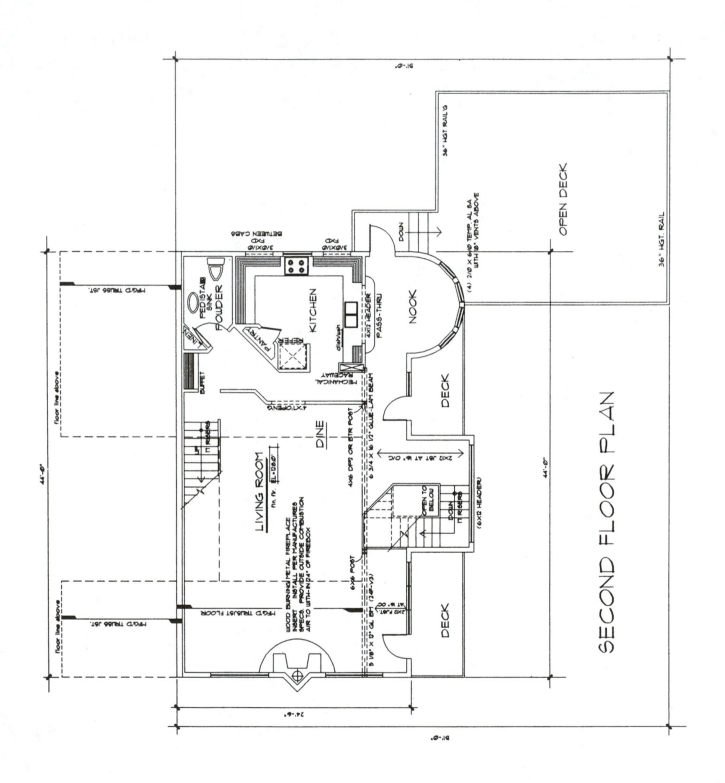

SECOND FLOOR PLAN

PROBLEM 10 (CONTINUED)

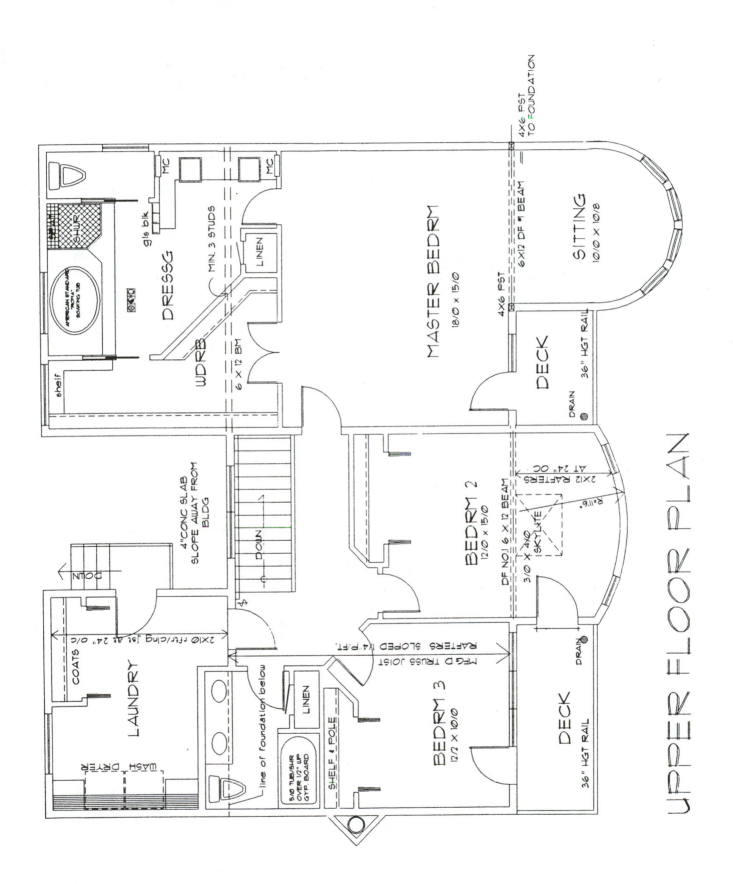

UPPER FLOOR PLAN